BIOLOGICAL AND CULTURAL DIVERSITY

The role of indigenous agricultural experimentation in development

Edited by
GORDON PRAIN, SAM FUJISAKA
and
MICHAEL D WARREN

Practical Action Publishing Ltd
27a Albert Street, Rugby, CV21 2SG, Warwickshire, UK
www.practicalactionpublishing.org

First published 1999\Digitised 2008

ISBN 10: 1 85339 443 2
ISBN 13: 9781853394430
ISBN Library Ebook: 9781780444574
Book DOI: http://dx.doi.org/10.3362/9781780444574

A catalogue record for this book is available from the British Library.

Since 1974, Practical Action Publishing has published and disseminated books and information in support of international development work throughout the world. Practical Action Publishing is a trading name of Practical Action Publishing Ltd (Company Reg. No. 1159018), the wholly owned publishing company of Practical Action. Practical Action Publishing trades only in support of its parent charity objectives and any profits are covenanted back to Practical Action (Charity Reg. No. 247257, Group VAT Registration No. 880 9924 76).

Typeset by J&L Composition Ltd, Filey, North Yorkshire, UK

This book is dedicated to the memory of Professor Michael Warren, whose untimely death occurred during the preparation of the volume. Mike's professional life was committed to the practical application of social sciences to development. His most powerful message was that in order to achieve successful development, we have to work with local people using their knowledge as the number one resource.

In implementing that message, he helped establish indigenous knowledge centres throughout the world and demonstrated very clearly the potency of local knowledge in understanding, domesticating and utilizing biological diversity. His interest in this theme was the driving force behind the present volume. Mike Warren's energy, commitment and eloquence in pursuit of people-centred development will be sadly missed.

Contents

PREFACE

WE WELCOME THIS important volume as the tenth in our series on Indigenous Knowledge and Development. While much attention has been paid to biological diversity, its close association with cultural diversity has been comparatively neglected: we hope that this volume will help to demonstrate the intimate relationship between the two.

As the title indicates, our 15 contributors all deal with farmer experimentation, illustrated with case-studies from many parts of the world. The crops that are used for experiments include rice, maize, beans, and root-crops; there are also studies of water harvesting (in the Sudan) and of home gardens (in South India).

The studies show that:

- farmers *do* experiment. In any society, there may be only a few individuals who are truly experimental, but this is true of all societies, including Western ones
- farmers know their local environments intimately, and their experiments are usually very site-specific
- farmers also have a close and detailed knowledge of local cultivars, and are well aware of the need to promote biological diversity
- the experiments underline the importance of *in situ* conservation, including the protection of wild plants
- any attempts at conservation of natural resources should involve the local inhabitants
- farmers' experiments are 'constantly being re-invented' – indigenous knowledge is a dynamic topic
- while contributors stress the value of farmers' experiments, they are also aware that these are no substitute for conventional on-farm research, but they do provide a valuable resource for the latter
- women farmers are often also actively involved in experiments.

This collection, with its detailed descriptions and analyses of specific and varied situations, should broaden the debate about biological diversity; it should also emphasize the importance of taking a broad approach, one that includes the cultural factors.

David W. Brokensha
January 1998

Notes on Contributors

BHUKTAN, JIT *Researcher, International Institute of Rural Reconstruction, Philippines.*

BROUWERS, JAN *Programme Officer, Netherlands Development Organisation in Cameroon.*

BUCKLES, DANIEL *International Development Research Centre, PO Box 8500, Ottawa, Canada.*

DENNING, GLENN *Director of Development, International Center for Research on Agroforestry, Nairobi, Kenya.*

FUJISAKA, SAM *Agricultural Anthropologist, CIAT, Colombia.*

NIEMEIJER, DAVID *Department of Irrigation and Soil and Water Conservation, Wageningen Agricultural University, The Netherlands.*

OU, LI *Deputy Director, Centre for Integrated Agricultural Development (CIAD), Chinese Agricultural University, Beijing, People's Republic of China.*

PERALES, H. *Ecologist, Colegio de Postgraduados, Montecillos, Mexico.*

PHILLIPS-HOWARD, KEVIN *Professor and Head of Geography, University of Transkei, South Africa until his untimely death in 1995.*

PINIERO, MARICEL *Doctoral candidate, Department of Anthropology and Linguistics, University of Georgia.*

PRAIN, GORDON D. *Regional Director, International Potato Centre, East, Southeast Asia and the Pacific Division, Jakarta, Indonesia.*

QAYUM, M. A. *Project co-ordinator, Action for World Solidarity, India.*

QUIROZ, CONSUELO *Director, Venezula Resource Center for Indigenous Knowledge (VERCIK), University of the Andes, Trujillo, Venezuela.*

RAJAN, VITHAL *Founder and Director, Deccan Development Society, India.*

RAJASEKARAN, BHAKTHAVATSALAM *Senior Information Scientist for Agriculture, Consortium for International Earth Science Information Network (CIESIN), Saginaw, Michigan, USA.*

ROLING, NIELS *Professor, Department of Communication and Innovation Studies, Wageningen Agricultural University, the Netherlands.*

SCHNEIDER, JURG *Anthropologist, Institut für Ethnologie, Universität Bern, Switzerland.*

STOLZENBACH, ARTHUR *Department of Extension Science, Wageningen Agricultural University, the Netherlands.*

TAYLOR, DAN *Director, Centre for Low Input Agricultural Research and Development, University of Zululand, KwaZulu, Natal, South Africa.*

WARREN, D. MICHAEL *Director, Center for Indigenous Knowledge for Agriculture and Rural Development (CIKARD) and University Professor, Iowa State University, Ames, Iowa, USA, until his untimely death in 1997.*

WARREN, MARY S. *Consultant in Rural Development and Gender Issues in Development.*

XIAOYUN, LI *Director, Centre for Integrated Agricultural Development (CIAD), Chinese Agricultural University, Beijing, People's Republic of China.*

ZHAOHU, LI *Lecturer, Agronomy Department, Chinese Agricultural University, and Deputy Director, Guioxin Association for Rural Technology Development, People's Republic of China.*

Introduction

DURING THE PAST decade development professionals have focused their attention on mechanisms that can facilitate participatory decision-making, collaboration with clientele groups, and sustainability of development project efforts (National Research Council, 1991). Since the 1980s the role of indigenous or community-based knowledge systems has been demonstrated as a powerful ally in the efforts to make development more clientele-friendly and more cost-effective. Indigenous knowledge is now viewed by many as the basis for indigenous approaches to decision-making, particularly through indigenous organizations where community problems can be identified and discussed, frequently leading to indigenous approaches to innovation and experimentation (see p.158 in Warren and Pinkston, 1997). In the past few years there has been an enormous increase in the number of publications on these topics, especially on indigenous experimentation (e.g., Adams and Slikkerveer, 1997; Amanor, 1991; Brush and Stabinsky, 1996; Leakey and Slikkerveer, 1991; Sillitoe, 1998, Thurston, 1992; Thurston, 1997; Thurston and Parker, 1995; Warren, 1994). Their efforts should expand with the new availability of user-friendly manuals and guides for recording indigenous knowledge (e.g. Center for Traditional Knowledge, 1997; IIRR, 1996). Global concern with the issue of biodiversity is also now being matched with a concomitant concern for the human knowledge generated in every community that is related to the biological realm (National Research Council, 1992).

Interest in the role of indigenous agricultural experimentation in fostering biological diversity and the cultural knowledge of that diversity has likewise expanded rapidly in recent years. An example of the interest is reflected in the International Conference on Creativity and Innovation at the Grassroots, co-ordinated by Anil Gupta and held at the Centre for Management in Agriculture, Indian Institute of Management, Ahmedabad, India in January 1997 with dozens of papers presented by participants from all over the globe. This volume provides additional case-studies that show how indigenous knowledge, practice, and experimentation can and do serve to advance agricultural development. This introduction places the current efforts to record examples of indigenous agricultural experimentation into a historical context, with a bibliographical section that reflects the rapidly growing volume of current publications on this subject.

That farmers work actively to solve their problems and do so based on locally developed knowledge are not new ideas. At the turn of the century King (1911) documented practices in east Asian rice-growing countries which formed the bases of many 'modern' rice-production technologies. Sauer (1969) encouraged building technical innovations on traditional knowledge in Mexico in the early 1940s. Cultural ecologists (or 'ecological anthropologists') documented farmers' practices and knowledge – albeit without much of a concern for experimentation and 'development' (Conklin, 1957; Evans-Pritchard, 1940; Forde, 1949; Leach, 1968; Netting, 1968; Rappaport, 1968; Steward, 1955).

A next wave of researchers documented cases of active and inventive indigenous agricultural problem-solving. Johnson's (1972) work with small farmers in Brazil, Bunch's appeal for farmer-centred agriculture (1985), and Richard's work in West African farming systems (1985) were among important earlier contributions. In addition to defining different types of indigenous experimentation, Rhoades and Bebbington (1995) reviewed some of the key contributions supporting the 'farmer as experimenter' written since Johnson's work more than 20 years earlier (Biggs and Clay, 1981; Box, 1989; Denevan, 1983; Rhoades, 1987; Thompson, 1973).

Richards (1986) expressed the general problem in reference to African agriculture:

> Many commentators see the prevalence of 'primitive' cultivation practices as a major factor in present food-production difficulties. Permanent solutions to the famine problem, it is supposed, will depend on closing the 'gap' between science and a 'backward' peasantry, hence the stress placed on 'top-down' agricultural extension systems for the efficient delivery of 'modern' inputs to small-scale farmers. According to this viewpoint, the key to an agricultural revolution in Africa is technology transfer.
>
> This book challenges the assumptions of this approach. It argues that technology transfer is part of the problem not the solution. There is indeed a major gap between science and the peasantry, but this gap is conceptual not historical. Tropical agricultural science is out of touch with its clientele. The problem is not how to bridge the centuries and reach 'backward' farmers with 'modern' inputs and advice, but that these inputs and advice are in many cases totally inappropriate to African food-crop producers because they have been designed without reference to the problems, priorities, and interests of those who are supposed to use them (1986: l).

Richards further discussed the need to mobilize local initiatives in solving problems given farmers' interest in, commitment to, and successes in technological change:

> Careful analysis of peasant innovation and experimentation... provides clues as to the nature of local priorities and the scope and trajectory of endogenous agricultural change. Rather than concentrate on technology transfer, formal-sector R&D should lock on to local initiatives. Scientific research should be carried out within design parameters set by the farming community... Formal-sector research agencies look at land and land resources from an analytical perspective... By contrast, farmers view the landscape in synthetic terms (1986: 156).

Agricultural research began to incorporate indigenous knowledge and farmer participatory research in the 1980s. Studies of indigenous knowledge systems and development were presented in a landmark collection (Brokensha *et al.*, 1980) and in a follow-up (Warren *et al.*, 1995). International agricultural research increasingly built research upon indigenous knowledge throughout the same period, with findings published in collections and reviews (Dvorak, 1993; McCorkle, 1994; Moock and Rhoades, 1992; Warren, 1991). In off-farm case-studies of innovations and experimentation (e.g. Adegboye and Akinwumi, 1990; Gamser *et al.*, 1990) demonstrate that 'the poor are experimenting constantly, innovating in a struggle to survive' (ibid.: ix) and that people's technologies 'develop in a way that retains and builds upon local skills and closely reflects the priorities of local people' (ibid.: x).

Many of these recent studies have described farmer experimentation and innovation. In a USAID-funded project in Niger:

> '...farmers were observed to manipulate such research-related concepts and techniques as the functional equivalent of a (necessarily oral) literature review before mounting field trials... , split plots and trial replications, experimental versus non-experimental variables, cost/benefit and risk assessments; team-like analyses of results, and hypothesis generation for future research' (McCorkle and McClure, 1995: 327).

Other studies have had more of an ethnographic basis, focusing on indigenous concepts and terms. Work with the Ara of Nigeria indicated that 'Management and development planning concepts often exist in local language... Local leaders and citizens have a comprehensive understanding of weaknesses and strengths in their own indigenous organizations and are very open to experimenting with new management approaches' (Warren *et al.*, 1996: 48).

In a critique of farming systems research, den Biggelaar (1991) noted that 'little has been done to develop indigenous technology generating and diffusing capacities already present in the rural areas.' He proposed a model:

> ...based on co-operation and collaboration between the exogenous and indigenous knowledge systems leading to a synthesis of the two. The underlying principle of the model is that the ultimate solution for rural development is not the dumping of more scientists upon rural people (of whatever discipline) to make exogenously-generated technologies more adaptable and in-line with people's problems, but to strengthen, empower, and legitimize indigenous capacities for identifying problems and developing solutions for these problems (1991: 25).

Indigenous capabilities need not be limited to farmers in developing countries. Jules Pretty (1991) noted that for the UK:

> Over a period of two centuries crop and livestock production increased 3–4 fold as innovative technologies and techniques developed by farmers were extended to other farmers through tours, farmer groups, open days, and publications, and then adapted to local conditions by rigorous experimentation. These technologies maximized the use of on-farm resources at a time when there was no government ministry of agriculture, no research stations, and no extension institutions (132).

Farmer participatory research – which builds on farmer experimentation and indigenous knowledge – has become the norm, generating a huge amount of literature (Ashby *et al.*, 1989, 1996; Berg, 1996; Biggs, 1980; Brouwers, 1993; Bunch, 1989; Bunch, 1990; Chambers, 1997; Chambers *et al.*, 1989; den Biggelaar and Hart, 1996; Hiemstra *et al.*, 1992; Innis, 1997; Joshi and Witcombe, 1996; McCorkle and Bazalar, 1996; Matlon, 1988; Page and Richards, 1977; Rhoades and Booth, 1982; Richards, 1985, 1986; Scarborough *et al.*, 1997; Selener *et al.*, 1997; Sthapit *et al.*, 1996; Systemwide, 1997; Tripp, 1989; Van Veldhuizen *et al.*, 1997; 1998; Warren, 1989; Willcocks and Gichuki, 1996; Witcombe, 1996; Witcombe *et al.*, 1996), reviews (Amanor, 1989; Bentley, 1994; Farrington and Martin, 1987), and catchy phrases such as 'farmer-back-to-farmer' (Rhoades and Booth, 1982), 'putting the first last' (Chambers, 1983), 'farmer first' (Chambers *et al.*, 1989), 'joining farmers' experiments' (Haverkort *et al.*, 1991), 'farming for the future' and 'low-external-input and sustainable agriculture' (Reijntjes *et al.* 1992), and, more recently, 'beyond farmer first' (Scoones and Thompson, 1994).

These contributions have led to the shared vision that, 'If we are to be serious about the development of a sustainable agriculture, it is critical that local knowledge and skills in experimentation are brought to bear on the processes of research' (Pretty, 1995: 180).

Farmer experimentation and knowledge is increasingly acknowledged as a key component of agrobiodiversity conservation and crop genetic-improvement. A recent collection of case-studies addressed 'the need to develop appropriate research and development strategies which build upon both the capacities of farmers to experiment with crops, and the knowledge they have acquired of diversity... Farmer experiments with crops have been important in promoting diversity and the conservation of species and varieties' (de Boef *et al.*, 1993: 1).

A more recent collection of experiences on participatory plant breeding (Eyzaguirre and Iwanaga, 1996) covers a continuum from breeder-controlled to farmer-controlled systems of breeding and selection, and indicates clearly the cost-effectiveness of working with farmers' skills in experimentation and innovation.

Sumberg and Okali (1997) have provided the most recent and very comprehensive overview of the literature dealing with indigenous agricultural experimentation and innovation. The authors also present empirical data on farmers' experiments from case-studies in Kenya, Zimbabwe, Ghana, and the UK and conclude that:

> ...farmers' experimentation is widespread, an important part of everyday farming, and shares many characteristics with formal agronomic experimentation. Thus, through their experiments, farmers are involved in ongoing processes of local knowledge creation through site-specific learning, which, in the short term, results primarily in small adaptations to farming practice and, in the long term, contributes to the development of new farming systems. However, we conclude at the same time that although in many situations the arguments for greater participation of farmers in agricultural research are compelling and relevant, relatively little potential synergy will be realized through formal research and farmers' experimentation being more closely linked. In addition, because of the site-specific nature of the knowledge created through farmers' experiments, the claim that there is significant unrealized development potential associated with them, which could somehow be used to make an impact on a larger scale, is also called into question (1997: 7–8).

> Both farmers' experiments and much formal experimentation aim to develop practical solutions to immediate problems or to seek small gains within the context of proven production methods and systems. Both are largely empirical and iterative, combining experience, observation (both methodical and opportunistic), intuition, persistence, skill and luck (1997: 149).

Bunders *et al.* (1997) presented a collection of case-studies looking at the application of both indigenous biotechnology and science-based biotechnology to agricultural development, building on farmers' knowledge while making use of the latest scientific insights. The case-studies deal with biotechnological practices of farmers in animal health, biopesticides, food processing, and crop genetic resources. A section on 'science-based biotechnology' assesses its potential within the socio-political context: while a section on 'building on farmers' practice' calls for an integrated approach to biotechnology development that builds on farmers' knowledge and makes use of the latest scientific insights. Similar topics had been discussed at a USAID-funded workshop on the role of indigenous agricultural knowledge in biotechnology that was conducted at Obafemi Awolowo University in Ile-Ife, Nigeria in 1996 (see Warren, 1996).

The 15 contributions to this volume span the globe, with case studies from Africa (Nigeria and one each for Sudan, Benin, Mali, and South Africa), Asia and the Pacific (Philippines, Papua New Guinea, Indonesia, India, China, and Nepal), Latin America (Mexico, Peru, Venezuela), Europe (Netherlands, France), and Australia. The authors come from Canada, the USA, UK, the Netherlands, India, Venezuela, South Africa, Nigeria, China, Nepal, and the Philippines. Most of the studies deal with various types of on-farm experimentation related to production agriculture and biodiversity, such as the development of new crop varieties, manipulation of germplasm, testing of exotic inputs, and pest management. Related studies focus on the dynamic nature of indigenous organizations in China and Nigeria, the clash of systems in South Africa between small-scale farmers and conservationists, indigenous concepts of experimentation (*shifleli*) in Mali, and water resource management in the Sudan. Of particular interest is the role of indigenous approaches to biotechnology including farmer expertise in the conservation, evaluation, and exploitation of plant genetic diversity (Prain and Bagalanon 1994). It is clear from this collection of case-studies that farmers world-wide are involved in experimenting, recreating and reinventing, a capacity that must be harnessed in participatory technology development.

1. Rice cropping practices in Nepal: indigenous adaptation to adverse and difficult environments

JIT BHUKTAN, GLENN DENNING, and SAM FUJISAKA

Abstract

NEPALESE SMALL FARMERS employ a wide range of indigenous practices and corresponding technical knowledge in organizing rice farming. Practices are well adapted to fragmented land and other diverse and difficult environments, while technical knowledge has been developed over years. Farmers conceptualize farm problems and design experimental processes for whole systems, and superimpose experiments on regular farming activities, to adjust the systems according to the changing environs. Knowledge of the indigenous practices can be invaluable in the design and conduct of rice research for increasing system productivity or maintaining sustainability.

Introduction

Rice occupies 27 million ha of 54 per cent of Nepal's total cultivated area (2.65 million ha). Mean yields have been 2.1 MT/ha. Between 1961–62 and 1988–89 rice production increased at the rate of 0. 41 MT annually, whereas population grew from 9.4 to 17.13 millions from 1961 to 1986. As a result, per capita rice availability decreased from 224 kg in 1961 to 138 kg in 1986 (DFAMS, 1990). Attempts made to improve rice production have not been based adequately on the traditionally inherited local farming practices.

Agricultural research in diverse and difficult environments has increasingly relied on the understanding of farmers' traditional practices and knowledge as a practical starting point for research intended to increase productivity and maintain sustainability (Brokensha *et al.*, 1980; Rhoades and Booth, 1982; Richards, 1985; Thurston, 1990; Warren and Cashman, 1988). Researchers have elsewhere incorporated farmers' perspectives and worked to develop farmer participatory research as means to make rice innovations farmer-appropriate and adoptable (Maurya *et al.*, 1988; Osborn, 1990; Prakah-Asante *et al.*, 1984; Richards, 1986, 1987). Such an approach is crucial for countries like Nepal, which is resource-poor and where rice is grown in diverse and difficult environments. This chapter examines local rice cropping practices of Nepalese small farmers, including their rice-growing environments and indigenous experiments.

Method

Descriptive data were gathered via 26 in-depth household interviews and via interviews, observations and group discussions with 190 rice-farming smallholders from 17 rural communities located at various altitudes of mid-Hill and Tarai rural communities of central Nepal. The data were analysed using simple statistics, and findings are presented in simple frequencies, descriptions, farmers' statements and narratives.

Results

The results are organized in four sections: rice farming environments, rice cropping practices, farmers' experiments, and farmers' varietal organization strategies.

Rice farming environments

Rice farming environments in the study area were characterized by variable altitude and diverse bio-physical conditions. The altitude in the Hill ranged from 525 to 1710 metres and in the Tarai from 60 to 185 metres. The Hill average monthly temperature ranged from 2°C to 27°C, and that of the Tarai 15°C to 30°C. Minimum temperature went down to ¯0.15°C in the Hills and 8°C in the Tarai and the maximum went up to 32°C and 37°C, respectively. The mean annual rainfall in the Hills and Tarai was 2141 mm and 1786 mm, respectively, with most prevailing during June to September. The minimum rainfall prevailed from November to December in the Hills, and from October to April in the Tarai. Rice farming environments in the Hills and the Tarai were different, while fragmented land was the most pervasive environment in rice farming.

Land fragmentation. The 91 Hill and 99 Tarai rice farmers cultivated 0.48 ha and 0.67 ha which were fragmented into three and five parcels, and scattered at a mean distance of 1.02 km and 0.57 km from home, respectively (Table 1.1). The distance was more of a problem in the Hills, where many rugged, steep-sloping narrow trails were passable only on foot. Hill ricefields were often located on river terraces, in basins or valleys; while other parcels and homestead were scattered over the Hills. Some Tarai parcels were located in other villages,

Table 1.1. Land fragmentation in the Hills and Tarai of Nepal

	Hills	*Tarai*	*Total*
Number of sample households	105	99	204
Mean size of total landholding (ha)	0.7	0.8	0.8
Mean number of parcels	5 (1–12)	4 (1–13)	5
percentage households with 1 parcel	3	7	5
2–4 parcels	39	47	42
5–7 parcels	41	25	32
8–10 parcels	16	18	17
10+ parcels	4	3	3
Mean size of parcels (ha):	0.14	0.24	0.19
percentage parcels of size up to 0.11 (ha)	0.1	43	20
0.11–0.2	39	23	32
0.21–0.4	15	15	15
0.4+	3	20	11
Mean distance: home to parcels (km)	0.9	0.5	0.7
percentage parcels w/distance (km)			
0.0–0.5	33	45	39
0.51–1.5	55	49	52
1.51–2.5	4	3	3
2.5+	8	3	6

although these were usually accessible by oxen-carts and bicycles. Parcel characteristics influenced farming practices.

Land forms. In the Tarai – where the relative elevation of flat land was of key importance – parcels were categorized as lowland (43 per cent of areas), midland (23 per cent), upperland (17 per cent), and deepland (9 per cent); (Table 1.2).

Hill parcels were characterized by elevation, slope and relative width of the cultivable surface. Hill lands were first categorized as upper hill (including summit), intermediate hill (*pahad*), foothill (*fedi*), or river basin (*byansi*). Each of these was further characterized as level (favourable for transplanted rice (TPR)) or sloping (less favourable – good for only upland direct-seeded rice (DSR)). Level and sloping terraces were further divided into wide and narrow: an oxen pair could be turned around easily on a wide terrace, while animal-drawn implements had to be unmounted for turning animals around to avoid their falling down the terrace wall on narrow terraces. The narrower and steeper the slope, the harder it was to cultivate. Overall, Hill rice parcel areas were 53 per cent narrow-levelled terrace, 31 per cent wide-levelled terrace, 10 per cent river valley, and 6 per cent Hill slope (Table 1.2).

Based on the land type, farmers had typified rice crops as: *pakhedhan* or *ghaiya*, ('rice grown on unbunded upland slope'); and as *khet dhan* ('rice in bunded levelled land'). They had further categorized *khet dhan* into: *kanle dhan* ('bunded terrace rice'), *byansi dhan* ('river basin rice'), *madi dhan* ('interhill valley rice'); and the plain Tarai rice as *madhesiya/deshi dhan*. The Tarai farmers

Table 1.2. Land types on Hills and Tarai parcels; per cent of area and percentage of farmers having different land types

	Hills		*Tarai*	
Parcel land type	*% Area*	*% Farmers*	*% Area*	*% Farmers*
Flatlands				
Upper	–	–	17	100
Middle	–	–	23	93
Lower	–	–	43	100
Deep	–	–	9	88
River valley or basin	3	20	2	18
Alluvial fan				
Flat terrace	4	43	6	21
Sloping terrace	3	50	–	–
Hill terrace				
Narrow, level	47	100	–	–
Wide, level	22	92	–	–
Narrow, sloping	9	84	–	–
Wide, sloping	6	70	–	–
Hill slopes				
Undulated, sloping	3	58	–	–
Moderate slope	2	43	–	–
Steep slope	1	32	–	–

had categorized rice as: *dih* or *tand dhan* ('upperland rice'), and *khala dhan* or *ghol dhan* ('lowland rice').

Soil types. Farmers first categorized soils into textural types (Table 1.3). The heaviest soil was sticky, suitable for potters, and hard to prepare. Hard clods emerged during tillage and could injure draft animals and damage implements, and crack when dry. Lighter soils were easier to plough and weed – but, compared with heavier soils, considerably less productive and costly in terms of irrigation and nutrient use. Aside from being prone to erosion, drought and wind damage, light soils also harboured more crop pests. Intermediate soils – a mixture of heavy and light soils – were considered to be favourable for growing rice safely.

More of the Hill land (64 per cent) compared with the Tarai rice land (40 per cent) contained very light to light soils. More of the Tarai (45 per cent) compared with Hill lands (23 per cent) contained heavy to very heavy soils. The Tarai dominated (24 per cent) Hill lands (13 per cent) in terms of area of intermediate soils.

Parcels of diverse soil textures were typical of all rice lands; and these characteristics were used to describe parcels: e.g., *chimto* or *chikat khet* (sticky soil rice field), *balu* or *balaute* or *ret khet* (sandy field), and *domatiya khet* (mixed loam field).

Local terms also designated soil fertility or productivity (Table 1.3). Only small areas of Hill (4 per cent) and Tarai lands (13 per cent) were classed as highly productive. More Tarai land (67 per cent) was average to moderately productive compared with Hill land (56 per cent). Conversely, more Hill (40 per cent) than Tarai land (20 per cent) was of slightly low to low productivity.

Parcel categories for fertility and productivity included *malilo* or *urbara* (fertile), *falne* (productive), *jieunndo* (living and self-restorative), *jod rasilo* (strong, nutrient-rich), *bieunjho* or *jinda* (nutrient-responsive), *dami* (precious), and *honahar* (promising). A parcel of low soil-fertility was named as *rukho* or *beubjah* (less fertile), or *kamjor* or *kamsal* (poor or weak).

Table 1.3. Soil texture and soil productivity on Hills and Tarai parcels; percentage of area and percentage of farmers having different soil textures and productivity

		Hills		*Tarai*	
Parcel soils		*Area*	*Farmers*	*Area*	*Farmers*
Texture:	Very light	27	92	9	88
	Light	37	100	31	100
	Intermediate	13	93	24	100
	Heavy	16	74	23	100
	Very heavy	7	53	12	91
Productivity:	Highly productive	4	34	13	63
	Mod. productive	18	86	35	100
	Av. productivity	38	100	32	100
	Slightly low prod.	29	100	16	100
	Low productivity	11	97	4	92

Moisture condition. Parcels were also categorized in terms of water sources: sky, surface, and underground. Rain was the main sky source of moisture; and sky moisture farming was known as *akasey kheti*. Surface water included rivers, lakes, and ponds. Seasonal springs (*mool* or *ghol*), dugwells and tubewells were underground water sources.

Parcels were further categorized by excess or deficient moisture conditions: e.g., excess moisture parcels were known as *moolkhet baguwa* or *dubaha* ('slow running surface water in all seasons'), *sim* or *itahawa* ('stagnant surface water in all seasons'), *dhapailo* or *dhap* ('swampy land, saturated sticky soil with water below'), *simailo* or *khala* ('wet land with some visible surface water'), *bhal khet* ('land flooded in ordinary rain'), and *chiso, oshilo*, or *oshaniya* ('wetland without visible surface water'). Similarly, moisture-deficient parcels were called *khadedi bari* or *sukhaha* ('drought-prone land').

Although 84 per cent of the Hill and 75 per cent of Tarai farmers had access to irrigation (covering 56 per cent and 59 per cent of land area, respectively), irrigated parcels could suffer from either excess or deficient water. Such parcels were irrigated from different sources at times under different water arrangements, and suffered from different problems, including siltation, flooding, and water theft.

Bio-physical constraints. Parcels were subject to various physical cropping constraints. In Hills, 86 per cent of the farmers had parcels (representing 34 per cent of their lands) facing north or between two hills that were affected by shade (Table 1.4). Rice planting time was strictly time-bound in such parcels. All Hill and 37 per cent of the Tarai farmers had some lands with stony soils. Stony soils injured rice transplanters' hands and slowed down work because care had to be taken to protect tools. Local categories characterized stoniness as: *dhungyan khet* ('land with rock outcrop'), *chhadi khet* ('land with mixed pebble soils'), and *gegryan khet* ('rice land with stone chips mixed in the soil'). Seed rate, quantity of labour and draught animals, and method of tillage differed in rice farming in stony and stone safe parcels.

Drought affected 41 per cent of Hill and 37 per cent of Tarai land, of 70 per cent and 41 per cent of farmers, respectively. Methods of tillage, planting times, fertilizer use, and moisture-conserving practices differentiated rice farming in

Table 1.4. Problems on Hills and Tarai parcels; percentage of area and percentage of farmers facing problem

	Hill		*Tarai*	
Parcel problem	*% Area*	*% Farmers*	*% Area*	*% Farmers*
Shade	34	86	0	0
Stony soils	48	100	9	37
Drought	41	70	37	41
Flood	12	32	34	81
Weeds	19	58	41	74
Insects	51	82	73	100
Animals	21	49	11	29
Birds	61	43	68	87
Rodents	73	74	82	83

drought-prone land from normal farming. Floods affected 12 per cent of Hill and 34 per cent of Tarai lands and were experienced by 32 per cent and 81 per cent of farmers respectively.

Different parcels were subject to diverse biological hazards. About 20 per cent of Hill and 41 per cent of Tarai land were affected by serious weed problems. The extent and intensity of insect, snail, crab, bird, and mammal problems varied from parcel to parcel. Some parcels were named after such hazards, e.g., *gangate khet* ('crab-infested rice land'), or *bandare pato* ('monkey-prone land').

Socio-cultural conditions. Land parcels also differed with respect to socio-cultural attributes such as tenurial status and strategic location. All sampled farmers were landowners who did not own all of the land they each cultivated. Farmers also cultivated parcels under tenancy, share cropping, and contract farming arrangements. In the Hills 83 per cent and in the Tarai 76 per cent of the cultivated land were owned; while the rest was obtained from others (Table 1.5).

For tenanted parcels, owners did not pay a share of inputs – and tenants had legal use rights and paid 25 per cent of the yield of the summer rice as rent. Sharecropping was an informal arrangement for equal sharing of inputs and outputs between owner and sharecropper in which the tenant would not claim legal rights and the owner would not evict a sharecropper who used the land for more than one year. In contract farming, land was leased for not more than one year and for a fixed rent. The contract was renewable with a change of parcel to avoid the lessee from claiming tenancy rights after tilling for more than a year.

Tenure insecurity, share paid, and risk of loss increased from legal tenancy to sharecropping to contract farming. Except in tenancy, owners imposed several conditions for cultivating land, including restrictions on chemical use, types and number of crops, planting of fodder trees, and use of tractors. Because of variable tenurial incentives, farmers organized rice-farming activities differently on owned and other parcels.

Land parcels differed with respect to their strategic locations. Home parcels, and close to roads, irrigation canals, or public places had higher values. On some parcels located near temples, family deities, or drinking water sources, use of chemicals was prohibited. Parcels near a Shiva temple were not ploughed by oxen (the ox is the ride of Lord Shiva).

The high numbers, large distances, and small sizes of parcels of diverse environments had advantages and disadvantages. Scattered parcels were endowed with diverse resource opportunities for rice farming and also provided

Table 1.5. Tenure status of Hills and Tarai parcels; percentage area and percentage of households

	Hill		*Tarai*	
Tenure status	*% Area*	*% Farmers*	*% Area*	*% Farmers*
Owned	83	100	76	100
Tenancy	3	11	4	18
Sharecrop	11	51	16	49
Contract	3	21	4	22

an opportunity for spreading risks. Farm operations could be temporally spaced in parcels located at different locations, allowing for rotational use of scarce resources.

It was difficult, however, for farmers to manage numerous parcel environments in terms of appropriate varieties and cropping practices. Farmers reported that costs in terms of materials, power, and labour, were also higher for farming small scattered parcels, with extra time required for travel and transport. Different parcels were also prone to different pests; and synchronized planting of adjacent fields was particularly difficult. Crops in distantly located parcels were often not safe from birds, animals, and theft.

Parcels located within the service domains of different irrigation systems made it difficult to co-ordinate water-management activities for different parcels. For parcels located in distant villages, farmers might have to maintain labour and draught-exchange arrangements in those villages too. Having multiple parcels was also associated with boundary conflicts. Diverse parcel-environments required a respective range of farming practices, varieties and inputs.

Rice cropping practices

Given the prevailing farm environments, farmers had organized rice farming activities differently in the Hills and in the Tarai, the common features of which are described in this section.

Organization of rice cropping

Rice was grown in all irrigated, 17 per cent of the Hill and 81 per cent of the Tarai rainfed areas (Table 1.6). Rainfed rice was grown in 11 per cent and 36 per

Table 1.6. Average household land utilization for growing rice

Land utilization types	*Mountain (A=0.48Ha)*	*Tarai (A=0.67Ha)*
Average household farm size (Ha)	0.73	0.81
Rainfed area (%)	44	41
Irrigated area (%)	56	59
Average household riceland (Ha)	0.463	0.748
Rainfed (%)	11	36
Irrigated (%)	89	64
% Rice land/Total farm size	63	92
Annually rice grown area (Ha)	0.532	0.912
Rainfed (%)	10	33
Irrigated (%)	90	67
% Annual rice area/total rice land	115	122
% Annual RF rice area/total RF* rice land	100	111
% Annual IRG* rice area/Total IRG* rice land	117	128
% Rainfed rice area/Total rainfed land	17	81

*IRG = Irrigated, RF = Rainfed.

cent, and irrigated rice in 89 per cent and 64 per cent of the total rice area, in the Hills and in the Tarai, respectively.

Since farmers grew rice more than once in a year, the annual rice hectarage of the average Hill household was 115 per cent and that of the Tarai 122 per cent. Hill farmers grew rice in 100 per cent of their rainfed and 117 per cent of irrigated rice land, whereas, Tarai farmers grew in 111 per cent of rainfed and 128 per cent of irrigated rice land, annually. The practice of growing rainfed rice twice a year was found only in the Tarai. Only 17 per cent of rainfed land was used for growing rice in the Hills, as compared with 81 per cent in the Tarai.

Rainfed rice was locally called *akasey dhan* in the Hills and *deosarni* in the Tarai, and irrigated rice *kule dhan* in the Hills and *sinchit dhan* in the Tarai. In the Hills, rice planted in a stagnant water field was named as *simpani dhan*, that planted using spring water as *moolpani dhan*, and the rice planted by collecting rain flood was known as *bhal chhopi ropne dhan*.

Farmers grew rice as spring, summer and autumn crops in rainfed, and as spring and summer crops in irrigated land (Table 1.7). They grew rainfed spring rice in upland as DSR in the Hills, and TPR in wet fertile Tarai lowland, both as early crop.

Farmers grew rainfed summer rice as a single crop of TPR in the Hills, and as DSR as well as TPR in the Tarai. The rainfed autumn rice was a temporally extended summer crop typical of Tarai, grown as a double TPR (second crop) using earlier preserved seedlings.

Farmers grew irrigated spring rice as an early crop of TPR, and irrigated summer rice as a single crop, as well as a second crop. The single crop everywhere was grown as TPR but also as DSR in some Hill submerged land with running water.

The irrigated second crop of TPR, was grown using fresh seedlings in areas where a spring crop of TPR was grown earlier. However, in the Tarai it was also grown as double TPR using earlier preserved seedlings – as a summer crop when

Table 1.7. Percentages of household and rice area by types of rice crops

Types of rice crop		*Mountain* *HH=91*	*A=0.48Ha*	*Tarai* *HH=99*	*A=0.67Ha*
Rainfed:	Spring-Upland DSR*	12	1	–	–
	Early TPR*	–	–	12	3
	Summer-Wetland DSR	–	–	8	4
	Wetland TPR	20	10	69	22
	Autumn-Double TPR	–	–	12	3
Rainfed Total		28	11	89	31
Irrigated:	Spring-Early TPR	14	15	46	37
	Summer-Wetland DSR	3	1	–	–
	[Single] Wetland TPR	89	73	99	35
	Summer-Single TPR	14	15	29	33
	[2nd crop] Double TPR	–	–	26	4
Irrigated Total		89	103	75	109
Total		91	115	100	136

* DSR = Direct Seeded Rice; TPR = Transplanted Rice.

planted in August in irrigated areas, and as an autumn crop when planted in September to October in rainfed areas.

Farmers had typified rice cropping temporally based on season, but Hill farmers named after the planting season, and Tarai farmers named after the harvesting season, which reflected the seasonally critical aspects of rice farming in the two ecozones. Summer rice was known as *barkhe dhan* (planted in wet season) or *ashade dhan* (planted in June–July) in the Hills; and as *sarihan* or *agahan* (harvested from late autumn to mid-winter) in the Tarai.

Hill farmers named the spring TPR as *chaite dhan* (planted from mid-March to mid-April) or *hieunde dhan* (winter rice – when sown in February); whereas Tarai farmers named it as *osahan dhan* (harvested in wet season) or *bhadaiya dhan* (harvested from mid-August to mid-September).

The Tarai specific autumn rice was known as *katika* (maturing in November) or *sarihan* (late maturing up to early December but planted in the autumn). The local typology developed for traditional varieties, was also used for modern varieties, even if it did not precisely fit the latter.

Rice farming process

Rice farming involved a series of vertically integrated processual activities that comprised diverse resource-management and risk-adjustment indigenous strategies.

Rice land preparation

Hill farmers began land preparation with land cleaning by removing stones, creepers, weeds and trash to prevent injury to draught animals, tools breakage, and to raise plough efficiency. They cut nearby tree branches and bushes to prevent shade, and repaired broken parts and blocked animal burrows to reduce pests. Tarai farmers removed only some trash.

Farmers in the Hills trimmed a thin layer of rice terrace walls yearly by removing the spongy bulged layer of soil-mass formed by microbial activity. They had experienced that the soil layer was as rich as good animal compost but, if not removed, absorbed rain water during the rainy season, softened and loosened the inner soil of the wall, which ultimately led to slumping. Application of soils trimmed from the wall helped to maintain soil fertility in adjacent fields.

Farmers trimmed walls in the winter, when moisture was moderate, because it was too wet in the rainy season and too brittle in the spring to be trimmed safely. The winter dew and frost helped grow a thin cover of green algae to seal the walls after trimming, enabling them to withstand light winter rain. Subsequent grass-growth, protected walls from dry spring winds and later summer rains. Grasses – also protective – finally served as green forage in the busy wet season.

Farmers first prepared dry rainfed rice land thoroughly to puddle quickly after rainfall. Hill farmers cut soil once and did one to two soil turnings to expose soils adequately. Tarai farmers cut dry soil by four to seven iterative ploughs, gradually increasing plough depth to prevent implement breakage, prevent large clods (thereby excess draught stress and injuries to working animals), to reduce labour costs in clod beating and adequately to expose all productive layers of soils to solar radiation for raising soil fertility, to kill weed seeds by desiccation and let birds scavenge on insects at various stages of their life-cycle.

Puddling involved iterative ploughing, harrowing and levelling using oxen-drawn implements. Unlike the Tarai, the Hill rice land puddling also involved a manual hoeing of left-over parts and levelling the earlier oxen-puddled field by a male using a spade, and this operation, as well as the operator, was known as *bause.* In the boggy/swampy inter-Hill valley land, farmers puddled the soil by human trampling, as animals would get stuck into the deep sticky clay soils. They prepared fragile narrow rice terraces by hoeing with a spade to prevent terrace breakage. The Tarai farmers usually puddled the well-saturated lowland soils simply by laddering.

Farmers maintained bunds by trimming sides and putting over and pressing new soils for impounding water, and grew various types of pulse crops on the top. But some Tarai farmers maintained permanent bunds to grow grasses for animal feed, and believed that grass protected the bund well.

Rice sowing and planting

Farmers sowed rice by direct seeding (DSR), transplanting (TPR), and double transplanting, but earlier they tested seed viability and prepared seedlings for TPR.

Testing seed viability. Farmers selected seeds carefully but poor seed germination was still a common problem. Most farmers tested germination at 20–25 days before sowing so they could acquire new seed if necessary. Farmers first pressed seed between their palms and discarded the batch if a high proportion broke. Seed appearing intact underwent further tests. A small quantity of seed was sown. If up to 75 per cent germinated, seed was used, but at somewhat higher rates. If 50 per cent germinated, seed was partially replaced. Seed obtained elsewhere was sown at higher rates if germination was not tested. Farmers also soaked seeds and observed germination or placed seed in water; those that settled to the bottom were used, while those floating were discarded.

Seedling management. In cold hilly areas, DSR seedbeds were covered with a thin soil layer and smaller leafy branches of *Albizzia sp* to protect the seeds from birds, conserve soil moisture, enrich soil fertility, and maintain warmer temperatures for seed germination. In WSR seedling nurseries in cold areas and prior to seeding, farmers incorporated the leaves of *Cannabis sativa, Adhatoda vasica, Artimesa vulgaris* or *Wrightia antidysenterica.* Leaves improved soil fertility and raised soil temperatures for better germination after seeds were broadcast the following day. Mustard oil cake dust was applied over the seed beds to hasten growth and strengthen seedlings against breakage when pulled when they were about 5cm tall.

For early rice in cold Hills, farmers soaked seeds in saline water for 20 minutes (to wash away the waxy coat on the seed hull), and dried them in the shade with a cloth cover for six hours, and seeded over clean water beds with an oil cake dust layer. They changed water at 12-hourly intervals to avoid frost-borne cold after 15 days, when the heating effect of oil cake was over. Those not using salt treatment used sprouted seeds and incorporated various types of heat-producing herbal leaves in the seedbeds prior to seeding.

In Tarai rainfed heavy soils, farmers seeded dry seeds over the dry beds a few days before rainfall, without soil cover. They guarded open seedbeds against birds until they became green. Covered dry seeds in heavy soil would have poor germination, but some did it by using sprouted seeds. In rainfed light soils, the Tarai farmers raised seedlings in dry seedbeds by the covered method, whereas in the irrigated wet seedbeds they used sprouted seeds.

Rice sowing. Farmers had categorized rice by the method of sowing. They named the direct dry seeded upland rice as *ghaiya*, wetland direct dry seeded rice as *chharuwa*, and the wetland direct wet seeded (soaked seed) rice as *jarai dhan*. Hill farmers named transplanted rice as *ropuwa* in the Hills and *datura* in the Tarai, and the Tarai specific double transplanted rice as *khaduwan datura dhan*.

Farmers sowed upland rice (*ghaiya*) by seeding dry seeds behind the plough by thinly intercropping maize to reduce risk. They sowed *chharuwa* by broadcasting dry seeds in swampy lands with running spring water (in some depressed inter-hill valley lands) over the dark thin layer of mustard oil cake applied earlier, so that seeds were visible by colour contrast, which helped uniform distribution of seeds. Also, because seeds stuck to the oil cake layer and were not washed away by running water, the oil cake hastened seed germination. *Chharuwa* was also used in excess rain flooded Tarai *ghol* land for similar reasons.

In the cold water-submerged land of Hills, and such Tarai lowlands, farmers directly seeded a mixture of sprouted seeds of two varieties (*jarai/jaruwa*) with very high seed rate (as a safety measure against crabs).

Tarai farmers directly seeded dry seeds when the earlier TPR or DSR crop died of drought, rain came after seedlings had overmatured, or seeds failed to germinate earlier. However, direct seedling (sprouted seed) was a regular practice in some submergence-prone and seasonally waterlogged lands, where transplanting was both difficult and time consuming.

Transplanting, with respect to the number of seedlings per hill and inter-hill space, varied by: type of growing space, time of planting, and tillering and rooting behaviour of the variety. Spatially, the soil texture and moisture condition, and prevalence of bio-hazards threat conditioned planting. Farmers planted fewer seedlings per hill and reduced the distance between hills, with an increase in elevation and decrease in moisture and vice versa, to minimize the intra-hill plant competition for nutrients. More seedlings per hill in the lower area also reduced the effects of crab damage, flooding and weeds. If the thickly planted rice plants grew unaffected, they were thinned to feed animals.

Farmers planting earlier or later than the peak season, used more seedlings per hill, and hills were spaced narrowly to maintain normal plant population after seedling mortality from off-season moisture and temperature stresses.

In general, farmers planted cultivars having vigorous tillering and rooting traits with fewer seedlings per hill, while they planted the lesser tillering or rooting cultivars more densely, to minimize the number of sterile tillers per hill. Farmers often cut the tops of tall seedlings (and fed them to animals), to prevent plant mortality from mud sealing their tops as a consequence of the plants bending down into the mud.

Double transplanting was a traditional practice in the Tarai, growing two rice crops of local late varieties within a wet season, using earlier preserved seedlings for the second crop, usually on boggy highly fertile lowland parcels. After the spring rice matured from late September to mid-October, fresh seedlings were not available for the second crop. Farmers therefore preserved the summer rice seedlings by planting thickly and densely in some smaller void space in June–July. They pulled these seedlings, which had already tillered, separated their tillers and then retransplanted. The technique was known as *khaduwan* and the crop matured at the same time as the other second crop planted in June–July.

Soil nutrient management

Farmers usually did not apply farmyard manure (FYM) to fertile, water-abundant and flood-prone rice lowlands, to avoid plant lodging; in flood-prone land to avoid risk of loss, and in distant parcels due to transport difficulties. Rice received FYM indirectly from that applied to preceding or succeeding crops. In general, farmers applied a moderate dose of FYM to less fertile light soils more in the Hills than in the Tarai. With lesser production of FYM and its increasing demand for other crops, many Tarai farmers applied FYM once in every two to three years in rotation to different rice parcels.

Farmers used a range of soil modifiers, including: FYM (100 per cent), crop stubble and weeds (100 per cent), soils trimmed from bunds or terrace walls (88 per cent), night soil (78 per cent), raw animal wastes (76 per cent), chemical fertilizers (74 per cent), ash (61 per cent), green manure (26 per cent), tree leaves and litter (48 per cent), oil cake (39 per cent), threshing waste (17 per cent), and topsoil from the forest (8 per cent). These were used in various combinations and over different land/soil types. Of 66 farmers using chemical fertilizers, 63 also used organic sources. According to one farmer:

> Animal manure is scarce; rain washed nutrients away every year; cold impedes nutrient uptake by crops; weeds rob nutrients; inorganic fertilizers spoil the soil; some crops suck excess nutrients. The use of several nutrient sources is a safety measure. Some raise soil temperature; some suppress weeds and help to decompose them; some disinfect the soil; while others help restore the soil. Soil is a lactating cow. It must be fed well to be able to milk it more and longer.

Water organizing

Although 84 per cent of the Hill and 75 per cent of the Tarai farmers had access to irrigation in 56 per cent and 49 per cent of their land area, respectively, rainfall greatly influenced rice farming. Farmers therefore, anticipated the time of rainfall using traditional cues, and organized rice farming accordingly. These included: when ants started carrying eggs it could rain within a fortnight; when ants and termites started making nests inside homes it could rain within a month; on cloudy days when salt started melting, chili capsules became flaccid and dry tobacco leaves limp, and it could rain in the near future; during April to May, if crows started collecting sticks, it could rain after a month.

Similarly, farmers anticipated the rhythm of rainfall: if an eagle lamented high in the sky, prolonged drought was most likely; if small flying termites (*chhichimira*) emerged from the ground at the onset of rain, heavy rain would continue for a week; and if large flying termites emerged during a mild shower a short-term drought would be inevitable.

Given the scattered rice parcels, farmers acquired water from multiple irrigation systems: free water from a common river or pond; private wells or springs; co-ownership – two households sharing a water source, group, communal and government (Table 1.8).

In most of the communal irrigation systems, water was distributed according to parcels, not by household, requiring each farmer to provide water almost every day. On occasion, the time at which water was available from different irrigation systems for different parcels, clashed. Farmers' water conservation strategies included: maintaining hard pan at plough depth to prevent percolation; adequate

Table 1.8. Percentage of households by type of irrigation system

Irrigation systems	*Hills*	*Tarai*
Free/common w/o Control	7	20
Private	13	69
Co-ownership of Two	8	–
Group	13	–
Communal	92	93
Government	–	46
Total households with irrigation	83	75

puddling to seal up pores to further minimize percolation; bunds being kept up and cropped, thus preventing bund crack and conserving water.

Crop protection

Some common rice pest preventive measures included: field flooding; soil exposure between tillage; maintenance of flowing water to prevent blast and suppress other soil-borne pests; use of pest preventive botanicals; destruction of insect dwelling places during bund and terrace upkeeping; cutting of nearby host plants; varietal rotation; synchronized planting; draining off water intermittently to suppress cutworms, and the like.

Although 57 per cent of farmers used chemical pesticides, traditional pest control practices were varied.

- A red leaf disease of rice was controlled using dry branches of a local pine tree, *Pinus sp* and the peduncle of jackfruit staked and submerged at the irrigation ditch inlets. Water was drained after three to five days and the field kept dry for two to four days. Farmers reported that chemicals released by the branches helped eliminate the disease, and moisture stress forced the emergence of new tillers and leaves.
- For a 'white' leaf disease, branches of *Vitex negundo* were staked in infested areas, water was allowed to stand for one to three days and was then drained for two to three days.
- For rice stem borer the hill farmers broadcast oil cake dust of *Bassia butyracea* after stagnating water and let the water dry in the soil. Some Tarai farmers broadcast *Azadirachta indica* seed dust into the mud to control rice stem borer.
- Farmers raised fowls to scavenge on insects, and birds attacking a heavily insect-infested crop were not driven away. Children were encouraged to collect and feed insects to chickens.

Weeds. Farmers had categorized rice weeds as pulling, hoeing, and cutting types. Pulling types were annual broad-leaf and aquatic weeds. Hoeing types were deeper-rooted grasses, suckers, vines or bulbous weeds. Cutting types were forage for animals, cut to help the crop, but not eliminated.

Weeds were also typified as digestible (*pachuwa*) or indigestible (*apachuwa*). Digestible weeds decomposed if pushed into soil, while indigestible weeds did not. Indigestible weeds were further divided into 'feedable' and 'non-feedable'. Digestible weeds were pulled and then pushed into the soil to decompose. Feedable indigestible weeds were collected for fodder. Non-feedable weeds were pulled and dumped in the manure pit or used as animal bedding, draped over the bund to dry and decompose.

The village messenger (*bisar*) system in the Tarai was modified and used for protecting rice crops against human theft and animal damage. The village messenger guarded the field in return for the harvest from the outermost two to three crop rows. If he was effective, his share was large. If the crop was damaged, the outermost rows were usually lost.

Harvesting, threshing and storage

Harvesting. Farmers harvested upland rice by picking panicle, and the plants that still remained green were gradually cut and fed to animals. They harvested summer rice generally by cutting plants close to the soil surface, mainly to maximize the quantity of straw. In the lowlands they cut plants at 25–35 cm to protect growing relay crops and to supply organic matter to the soils.

Drying and stacking harvests. Farmers dried rice harvests in the field for between three and seven days, to reduce moisture content and transport costs; to prevent grain spoilage during curing in the stack (which was locally known as *hakuwa* – discoloured and bad-smelling grains that had poor taste and cooking quality, and fetched low prices); and to raise the storability of straw, and milling recovery of grains. They collected harvests and bundled in the day after drying off morning dew, to prevent spoilage in the stack. Farmers stacked bundles circularly, with panicles to the inside, to cure for easy threshing and protecting the grain from adverse weather. The size of stack also helped farmers estimate the labour and animals required for threshing.

Threshing. Farmers threshed *ghaiya* panicle tops by stick beating and the early rice by bundle beating immediately after harvesting to protect grains from rain. They threshed summer rice twice in the Hills, first by bundle beating to make urgent grain wage payments to labour and power, to repay debt, and to collect whole straw for roofing and weaving. They did second threshing by animal trampling (*dain*) to separate the remaining grains, and to clean and soften straw for animal feed. Except for seed, Tarai farmers threshed rice only by animal trampling (*dauni*).

Storage. Farmers stored grains in all types of containers – clay pots, jute bags, metal vessels, wooden boxes, old cloth packets, bamboo and straw-mat bins. They often placed these near the fireplace to disinfect grains by kitchen smoke. Mixing grain with neem seed dust, ash, and a few neem or hemp leaves were traditional seed protection practices. Usually farmers kept a cat to protect stored grains against rats. They did not kill home lizards, since these prey on insect pests. Marigold flowers in the Hills and onion bulbs in the Tarai were used to repel the seasonal flying insect pests. Farmers checked stored grains and turned occasionally, and, if needed, dried them in the sun to improve storability and conserve seed viability.

Farmers' experiments

Farmers' rice farming practices were the product of generations of continuous indigenous research. They studied through observing the actual crop growth processes, changes in various vegetative and reproductive behaviours of crops by manipulating growth factors, and through mutual sharing of the knowledge so gained. Based on such experiences, they dynamically organized rice farming under diverse environmental conditions. Farmers conceptualized farm problems and experimental processes in terms of whole systems, and superimposed their experiments on regular farming activities. The following two cases will exemplify farmers' applied field experiments and their rationales.

Experiments on new rice crop and varieties. Farmers experiment with cultivars to fit local environments and allow modification of cropping patterns. Krishna, a local rice farmer, tested an early maturing (10–110 days) rice, CH45, as an early or spring crop. He conducted experiments on 0.04 ha which were irrigated by stream and spring sources. Krishna wanted to determine if the spring rice could be grown profitably without negatively affecting the summer rice.

In the first of four years of testing, Krishna irrigated the seedling nursery with spring water. Although germination was supposed to be in 4–5 days, it took 8 days and was not uniform. The variety apparently did not do well in muddy water because although the germ had broken the hulls, it had died at emergence. A 0.05 ha plot was prepared, manured, and transplanted in the last week of March. Seedlings were small and hard to pull, even at 35 days. Tillering was poor. The crop was harvested in the third week of June. Panicles were big; but most bore empty grains due to ricebug and poor grain development. Yield was only 0.3 t/ha – compared with 2.6 t/ha in the summer crop. Stream water had been used to irrigate the crop.

In the second year, Krishna tested muddy versus clear water in his seedling nursery beds, seed from the last season's crop versus newly acquired CH45 seed, and spring versus stream irrigation. Separate seedling nurseries were prepared, sown in mid-February, and irrigated with spring water. Seed was either sown immediately after puddling or when water was clear after a day. Seedlings germinated poorly in the muddy beds; but did better in clear water. Four plots were prepared, with FYM applied to each. Rice was transplanted in the last week of March. The four plots were transplanted to seedlings from home-grown or new seeds, and were irrigated by spring or stream water (resulting in four treatments). Seed source did not appear to make a difference. Plants were stunted and tillered less in stream-watered plots; while panicles were bigger and had more grain in spring-watered plots. The yield from stream-watered plots was 0.3–0.4 t/ha; while the yield from spring-watered plots was 0.6–0.7 t/ha. Krishna concluded that the colder stream water was bad for the early rice.

In the third year, two 0.05 ha plots were planted, with both watered by spring. In one plot, ammonium sulphate at 50 kg/ha was applied at 10 DAT. Plants turned a deep green, grew faster, and had better tillering than those on the other unfertilized plot. The fertilized plot yield was 1.5 t/ha; while the unfertilized plot yielded 1.2 t/ha.

In the fourth year, Krishna tested home-grown CH45 and a new variety, K39, acquired from a friend. K39 was reportedly better than CH45 in taste and had the same duration as CH45. Krishna also liked the grain size and shape of K39. He tried K39 with fertilizer using both stream and spring water. K39 yield was 2.3 t/ha in spring water and 1.1 t/ha in stream water. At the same time, the mean yield of CH45 was 1.5 t/ha. Krishna has dropped CH45 and now grows K39 as an early season rice in areas watered by spring. He is still looking for a variety to grow in the stream-irrigated area, and still faces the difficulty of seed storage during the wet season.

Farmers' varietal adaptive experiment. Another farmer, Purna, received about 100 grams of seed of each of four rice varieties from a relative – who, in turn, had obtained them from a government mini-kit testing package. Each variety was recommended for different conditions. Purna was interested in one recommended for cold tolerance. After raising seedlings, Purna transplanted a few hills of the supposedly cold-tolerant variety on to different parcels, including: (a) a colder shaded area at 1100 metres above sea level, (b) a rainfed field at 800 masl, and (c) a swampy, river basin area at 500 masl. Crop growth, tillering, and grain filling were poor at the high altitude site. Plant growth was good, but only a few tillers bore full grains in the middle plot. Growth was vigorous, tillering was good, and panicles were long and full in the lowest plot – where no rice cultivar had previously been as high-yielding. The so-called 'cold tolerant' variety proved to be most suitable for the lower altitude swampy land. Subsequently, Purna learned that the variety (BG-90) had vigorous roots, and sturdy tillers requiring ample water, and that his relative was mistaken in terms of the recommendations. Since then Purna has grown only BG-90 in the low area.

Farmers' varietal organization

Farmers conducted various other types of action-based experimental and non-experimental tests, of which their varietal tests were of special importance in the context of farm level varieties organization.

The 190 farmers in the two areas grew 56 different rice varieties, of which 75 per cent were traditional varieties (TVs), and 25 per cent were modern varieties (MVs). Out of the 42 TVs, 38 per cent were grown in the hill; farmers planted TVs on 64 per cent of the hill farmers' total household rice area; while 100 per cent of the Tarai farmer planted TVs on 46 per cent of the total Tarai household rice area (Table 1.9).

Out of the 14 MVs, 50 per cent were grown in the hills, 29 per cent only in the Tarai, and 21 per cent in both areas. MVs were grown by 59 per cent of the hill farmers on 36 per cent of their rice area, and by 93 per cent of the Tarai households on 54 per cent of their rice area (Table 1.10).

On average, hill farmers grew four and Tarai farmers grew five different rice varieties (Table 1.11). Over 90 per cent of the Hill and 100 per cent of the Tarai farmers grew between two and eight varieties. More Hill (41 per cent) compared with Tarai farmers (7 per cent) grew only TVs. None of the Tarai and 19 per cent of the hill farmers grew only MVs. More Tarai (93 per cent) than hill (40 per cent) farmers grew both TVs and MVs (Table 1.11). Only two TVs (Thankote and Pokhreli masino) were dominant in the hills; while no TVs dominated in the Tarai. And only one (Masuli) in the hills, and two (Masuli and CH45) in the Tarai were the dominant MVs planted.

Farmers' varietal tests, preferences and selection criteria

Farmers invariably examined various aspects of any new or alien variety, by test-growing the variety, during actual cultivation, and even by observing new varieties grown in other's fields. Some of the common rice varietal attributes that farmers tested are described as follows, and summarized in Table 1.12.

Table 1.9. Traditional rice cultivars grown in the Hills and Tarai, percentage of farmers and per cent of area

	Hills		Tarai	
Cultivar	*% Farmer*	*% Area*	*% Farmer*	*% Area*
Battisara	46	32	–	–
Pokhreli Masino	35	15	–	–
Thankote	14	2	–	–
Chandramarsi	12	2	–	–
Ghaiya	10	1	–	–
Hansaraj	8	1	–	–
Bardana	8	1	–	–
Setomarsi	7	1	24	5
Chhote	6	1	–	–
Amjhutte	6	1	–	–
Sikarimarshi	6	1	–	–
Ratokathe	6	1	–	–
Kalokathe	4	1	–	–
Gartinaine	3	1	–	–
Thademasino	2	1	–	–
Kalodhan	2	1	–	–
Thanti	2	1	–	–
Muturi	–	–	44	9
Anadi	–	–	24	3
Mansara	–	–	18	2
Basmati	–	–	16	1
Madhukar	–	–	16	1
Assamemasino	–	–	13	3
Gurdi	–	–	11	2
Katika	–	–	11	2
Kalanamak	–	–	11	1
Hakijhukan	–	–	10	1
Haribhakte	–	–	9	1
Kasturi	–	–	9	1
Assamejhinwa	–	–	8	2
Kanakjeera	–	–	8	1
Sanikharika	–	–	8	1
Barmabhusi	–	–	8	1
Gokulchand	–	–	7	1
Babani	–	–	6	1
Barchhabahar	–	–	6	1
Aga	–	–	6	1
Bachhi	–	–	6	1
Begani	–	–	6	1
Chinaburo	–	–	6	1
Gauria	–	–	6	1
Tulsiprasad	–	–	5	1

Grain type. Farmers examined several aspects of rice variety grains, for seed, domestic consumption, and market purposes. Although higher grain yield was the dominant concern, farmers also carefully observed yield consistency over time. For seed, farmers tested the variety for its germination behaviour. They

Table 1.10 Modern rice cultivars grown in the Hills and Tarai, percentage of farmers and percentage of area

	Hills		*Tarai*	
Cultivar	*% Farmers*	*% Area*	*% Farmers*	*% Area*
Masuli	37	11	72	33
CH45	19	3	58	11
Janaki	9	1	14	2
Mallika	3	1	11	1
BG 90	26	6	–	–
K-39	20	4	–	–
BG-424	20	1	–	–
IR-8	21	5	–	–
Himali	13	3	–	–
Durga	11	1	–	–
Sarju 49	–	–	11	3
Sabitri	–	–	11	2
Pankaj	–	–	11	1
IT	–	–	8	2

preferred the quick-sprouting type, but one that germinated too quickly was difficult to store safely as it sprouted even in brief contact with moisture. The late-germinating one could have high storability but carried a risk of poor germination, and some special care and techniques were required for timely germination, involving additional resources. Most farmers, therefore, rejected too-quickly and too-slowly germinating cultivars under normal conditions. Farmers also examined the thickness of hull and waxy coat, often by opening the grain or scratching the seed coat with their finger nails. The thicker the hull and waxy coat over it, the longer it took to germinate, and it required adequate moisture and warmth for timely germination. The Hill farmers, especially in the cold area, therefore avoided such a variety.

Farmers had to store seed at least for six months and good grains for almost a year. Since grain spoilage during storage would mean seed and food shortage, farmers tested storability of the variety, and often avoided rice varieties with less storable grains.

Table 1.11. Number of rice varieties grown by farmers in the Hills and Tarai

Number of varieties	*Hills (%)*	*Tarai (%)*
1	10	–
2	20	4
3	34	17
4	18	28
5	11	18
6	2	20
7	3	7
8	2	6
mean number of varieties	4	5

Table 1.12 Percentages of households by type of attributes of new rice varieties they tested in the past

Attributes of new rice varieties tested		*Household % (N = 190)*
Yield:	Yield quantity	100
	Yield consistency	100
Grain type:	Grain colour	74
	Grain size and shape	78
Seed traits:	Seed germinability (proportion)	100
	Germination period	100
	Seedling maturity period/strength	100
	Transplanting time	100
	Seed storability/preservability	100
Plant type:	Plant height	100
	Maturity period	100
	Tillering vigour (numerous vs. few)	92
	Tillering habit (gradual vs. simultaneous)	89
	Proportion to grain bearing tillers	65
	Rooting habit	100
	Lodging habit	98
	Susceptibility to cold and shade	41
	Susceptibility insects/disease pests	100
	Threshability and shatterability	99
Growth factors:	Manure/fertilizer requirements	100
	Water requirement	100
Market quality:	Grain weight/volume	87
	Market price	53
	Milling recovery	94
	Grain storability (also for food)	86
	Weight loss during storage	61
Food quality:	Cooking quality (volume increase when cooked)	100
	Heaviness of cooked food (hunger satiability)	100
	Stickiness	64
Feed quality:	Palatability of straw to animals	100
	Straw storability	68
Others:	Durability of straw as roofing material	54

Farmers growing rice for market and selling by weight, tested grain-weight-loss during storage. They tried to prevent weight loss by adequate drying of harvests, but discarded those with an inherent weight-loss trait. Farmers also examined the threshability versus shatterability of rice varieties. Poor threshability meant loss of grain in the straw, and more threshing cost in terms of labour and animal power. However, shattering or falling of grains from panicles at maturity in the field and during harvesting, harvest drying and transporting also resulted into substantial grain loss. Farmers, therefore, preferred those that could easily be threshed but which did not shatter.

Farmers also, at times, examined the shape and size of the grain. The finer the grain with bright white kernel, the more attractive it was for market and the higher the price it commanded. Women considered a bright brown hull colour as advantageous for cleaning the grains conveniently, as they could easily detect foreign materials mixed with the grains.

The higher milling recovery of grain was important both for food and market purposes. Farmers discarded rice varieties with poor milling turn-out for not fetching better prices and having poor cooking quality.

Cooking quality of grain being important for both domestic consumption and the market, farmers tested these varietal traits. Poor cooking quality meant loss of grain, thereby requiring more rice to feed fewer persons – and it was also a sticky rice. They preferred a rice variety for food which increased in volume when cooked and was heavy when eaten, so that a smaller quantity of rice satisfied the hunger of more persons for longer. Farmers also believed that heavier and non-sticky rice food was more palatable and nutritious.

Plant type. Farmers examined various aspects of plant type of rice for various purposes. They required fairly tall plants, mainly to harvest more straw for animal feed, mat weaving (for human beds and seats), making tying ropes, roofing and fence materials. Farmers also stored and fed it to animals almost throughout the year. Farmers tested the lodging habit of rice plants. A lodged crop meant loss of grain and more labour cost to harvest it. They therefore preferred shorter varieties to the lodging taller ones.

Farmers carefully examined the tillering vigour of rice varieties, especially the proportion of productive tillers. They had experienced in the past that crops with a higher proportion of unproductive tillers not only yielded less grains but also produced straw which was less storable and palatable for animals, and less durable for roofing and weaving mats. As the unproductive tillers got spoiled faster, these reduced the storability of the bulk of the straw. Farmers preferred rice varieties with vigorous tillers, most of which were productive.

Farmers also examined the tillering habit and preferred one that tillered gradually to one tillering simultaneously. In the former type, even if the initial tillers were damaged by insects, the subsequent tillers had a chance to survive, or tillering could be induced through subjecting the crop to water stress conditions.

The crop maturity period was an important consideration, especially when they grew rice in relation to other crops. Farmers preferred a variety that matured at a definite time, even if planted late. One of the preferred attributes of Masuli was that even if its planting was delayed by two months it matured in early November together with others planted in July.

However, preference for the maturity period of a variety differed according to the environment, and the type of cropping pattern of which rice was a component. For the rainfed or water stress condition, and where moisture became scarce at the later part of the wet season, farmers preferred short-duration cultivars. Also they preferred early maturing varieties for the distant parcels, to harvest earlier to avoid theft, and they grew rice as an early crop or in between two crops, especially when delaying the establishment of the succeeding crop was not environmentally permissible.

In the land where water remained for a long time and other crops could not be grown thereafter, farmers preferred long duration varieties that matured after excess water problems and aquatic leeches had subsided. Under normal conditions farmers preferred medium-duration varieties because it was difficult to manage the too-short duration cultivars within a limited time period, and too-long duration ones upset the succeeding crops.

Farmers belonging to different ecological zones and having their rice land parcels scattered over different microecological pockets faced different types of insect and disease problems. They therefore preferred the rice cultivars that were resistant to local pests, and often discarded those which were susceptible.

All farmers tested to determine the water requirements of any new or alien variety. They had experienced that the less vigorous and shorter rooting of a rice plant had several disadvantages, including higher susceptibility to moisture stress and lodging. Likewise, plants with vigorous rooting required ample and running water, which farmers often lacked in most parts of their rice land. Farmers preferred a cultivar with relatively fewer and longer roots, for their light and moisture-stressed fields, and heavy rooted varieties for moisture-rich lowlands.

Farmers also did not like rice varieties requiring high levels of chemical fertilizers (despite their higher yields). Cash spending, unavailability of appropriate types of fertilizer, and more importantly, the adverse effects of chemicals in soils, had dissuaded small farmers from adopting such varieties. Farmers' varietal choices reflected the environmental diversity of their parcels. As an example, one Tarai farmer grew seven varieties on 0.7 ha and nine parcels. An early MV (CH45) was planted in the spring, followed by an MV (Masuli) in the summer on two irrigated, middle-light loam soil parcels. A short-duration TV (Mansara) was planted on a distant rainfed upper paddy. Mansara did well on the two irrigated parcels with heavy soils. Muturi was selected because it did not lodge due to excess water and high soil fertility. Muturi was followed by transplanted Masuli. A spring crop of Muturi was followed by a non-lodging tall TV (Hakijhukan) on a boggy, fertile lower parcel. A late-maturing TV (Katika) that required less water at later growth stages was planted on a heavy-soil, moderately fertile and partially irrigated parcel.

Different varieties were suited to different areas within the hills. Cold and shading were problems in the upper hills, making cold- and shade-tolerant TVs such as Thankote desirable. In shaded lands irrigated with warm spring water, shade-tolerant TVs such as Chandramarsi were grown. In sunnier irrigated areas of the mid-hills, some widely adaptable TVs such as Chandramarsi were grown. In sunnier irrigated areas of the mid-hills, some widely adaptable TVs such as Pokhreli Masino (see Table 1.13) were common. In the lower hills and river

Table 1.13. Farmers' (n = 48) reasons for growing Pokhreli Masino

	% Farmers
Cold (air and water) tolerant	100
High straw production	100
High grain quality & market price	100
Good storage	100
High milling recovery	100
Good disease resistance compared with other TVs	100
Does not require too much water	100
Lower seed rate than other varieties	100
Best for delicacies and ceremonial purposes	100
Compatible with following wheat crop	95
Does not require too much fertilizer but responds to moderate dose	88
High yield among LVs	83
Productive tillers	68
Full grains	34

basins with warmer climates and less shade, the MV Masuli was planted extensively. The short-duration MV, CH45, was adapted to light soils and drought; and the MV K-39 was adapted to fertile soils and drought. In areas with secure early irrigation and possible drought later in the season, the relatively early-maturing MV IR-8 was common. BG90 also grew in these areas; but IR-8 was planted because it matured earlier and was less affected by rice bug.

The diversity of rice cultivars also reflected farmers' multiple uses of rice, including consumption, fuel, livestock feed, crop production, a source of cash, medicines, roofing and handicraft materials, and liquor (Table 1.14). Farmers also grow some TVs as high-quality cash crops. These included Pokhreli Masino and Sikarimarshi in the Hills and Kalanamak, Kasturi, and Kanakjeera in the Tarai.

Farmers' rice cultivars changed over time as they routinely collected and tested varieties. Hill farmers adopted the TV, Thankote, in the 1960s. Thankote replaced several TVs (i.e., Ratemarsi, Chinia, Falame, Jungemarsi, and Beli); and now farmers were testing MV Himali to replace Thankote. Farmers in the low Hills are adopting a TV, Battisara, for its high yield, and it was likely to replace some current TVs. IR-8 and Masuli previously replaced several TVs, such as Rajbhog, Manbhog, Biramphool, and Baglunge.

Farmers in the Tarai were testing MVs such as Janaki, Laxmi, Sabitri, Mallika, and Bindeswori. Farmers had recently dropped some TVs such as Jogini, Sattari, Aga and Barmabhusi.

Farmers had increasingly sought varieties for early-planted crops. Hill farmers tried CH45 and K-39. These performed well with warmer spring water irrigation; but farmers were still seeking varieties adapted to the cold stream water.

Table 1.14. Farmers' (n = 37 in the Hills and 73 in the Tarai) reasons for planting Masuli

	Hills	*Tarai*
Adapted to various environments (except cold, shade)	100	100
Good grain quality and market price	100	100
High tillering and low seed rate	100	100
Good storage and germination	100	100
High straw yield and fodder quality	100	100
Does not shatter but easily threshed	100	100
Straw good for mats	100	100
Does not need running water	100	100
Volume increases upon cooking and is fairly heavy	100	100
Does not upset following crops	100	100
Responds to moderate amounts of fertilizer	100	82
High yield	83	95
High milling recovery	69	90
High tillering with few unproductive tillers	64	82
Heavy grain and minimum weight loss over time	59	85
Gradual tillering means less stem-borer damage	16	46
Roots not vigorous but some are deep, making it suitable for both light and heavy soil	3	28
Good for double transplanting	0	60

Two dominant varieties. Pokhreli masino covered 32 per cent of the hill farmers' rice area. Introduced in the late 1960s, it replaced many TVs. It was popular for a wide range of reasons (Table 1.15).

Masuli was the only widely adopted MV grown in the lower Hills and in the Tarai. It had replaced TVs since its adoption in the 1970s. Although farmers reported many favourable characteristics (Table 1.14), Masuli had reportedly started to deteriorate, as indicated by declining yield, increasing susceptibility to pests, reduced tillering and increasingly unproductive tillers. Farmers are looking for a replacement, and have rotated it among rice environments, readjusted inputs, and exchanged Masuli seed from household to household within and between communities.

Table 1.15. Farmers' (n = 91 in the Hills, 99 in the Tarai) uses of rice

	Hills	*Tarai*
Consumption		
Primary staple	10	95
Secondary staple	70	5
Livestock feed		
Straw	92	100
Rice hull, bran	79	100
Green forage	18	21
Fuel		
Straw mixed with dung cakes	–	91
Straw	26	94
Rice hulls	–	47
Crop production		
Seed	100	100
Straw converted to FYM	100	100
Residues incorporated	100	100
Straw used as mulch	27	38
Straw burned to enrich soil	21	11
Income generation		
Grain is sold for cash or bartered	79	100
Milled rice is sold/bartered	24	53
Straw is sold for cash or bartered	12	42
Milling by-products are sold	4	22
Miscellaneous		
Mats, baskets and ropes from straw	100	100
Foods from rice prepared for the sick	100	100
Ceremonial use	100	100
Poultry nests from straw	91	54
Straw used in pyre for cremating dead	71	94
Beer, liquor from grain	54	42
Roots used as animals medicine	37	21
Roofing material from straw	35	28
Wall & floor plaster from straw/hulls	35	89

Variety protective strategies
Some of the rice practices of farmers were strategies for sustaining the varietal viability. Among all crops, rice varieties were relatively short lived. Among others, seed selection and varietal rotation were the main strategies.

Seed selection. Farmers generally used their own home-grown seed. They selected seed to maintain the best of varietal qualities, and to prevent seed unavailability at the time of sowing, cash spending in buying seeds and hazards that often accompanied seeds from outside.

Farmers' seed selection practices included: selection and collection of heavy panicle-bearing matured, full and undamaged grains, tillers bearing heavy panicles, and whole plant cut from the best yielding patches – all from within the crop field; selection and collection of heavy panicle tops, tillers bearing heavy panicles, or harvest bundles containing heavy panicles before stacking the harvest (seed viability was believed to deteriorate during curing in the stack). The first type was more common in the Hills and the other one in the Tarai.

Varietal rotation. Rice varietal combination patterns were not permanent arrangements of rice crop in relation to other crops, even within the viable age of any rice variety. Farmers believed that continuously growing a rice variety in the same land, year after year, would bring about the early varietal degradation, in terms of declined grain and straw yields, increased succeptibility to insect and disease pests, and distortions in other traits. Unlike in the case of other crops, in which the crop itself was rotated over space and time, the varieties were rotated in rice farming, and was locally known as *dhan ko nal ferne.*

The small landholding size with diverse micro-environment did not usually permit a planned rice varietal rotation over space, which was otherwise a common strategy among the medium-sized and large holders. Stabilization of rice yield was, however, more crucial for small farmers, and spatial and temporal variety rotation was a strategy to attain it.

Spatially, farmers changed growing-space variety after three to five years, by growing it in other parcels with near suitable environments for it. Whenever this was not environmentally possible farmers exchanged the variety (seed) for another. The change in growing space often restored some of the degraded qualities, if not all. Farmers re-exchanged the variety after growing it for a few years. When the inter-farm seed exchange within the community ceased to show desirable improvements, farmers exchanged the seed with those from distant communities, and if still no sign of improvement was seen, it was finally replaced.

However, together with, prior to, or after the spatial rotations, farmers also rotated rice varieties temporally. Farmers often used different rice varieties for the spring and the summer rice, but they also used some varieties grown earlier as spring crop the next year as summer crop, and vice versa. Also, farmers dropped some varieties for a few years and grew them again thereafter by acquiring the seed from other households.

Conclusion and implications

Nepalese small farmers employ a wide range of traditional practices and corresponding technical knowledge in organizing rice farming. Practices are well adapted to fragmented land and other diverse and difficult environments, while technical knowledge has been developed over generations.

Household rice land was a cluster of multiple heterogeneous smaller parcels scattered over diverse ecological pockets. The individual parcels were not only different with respect to size and location, but also in terms of vital farming environmental attributes. Each of the individual parcels had a unique ecosystem.

Overall, farmers managed an aggregate of multiple agroecosystems. Ecosystems characterization procedures need to consider land fragmentation and the parcel-specific nature of farmers' adaptive strategies. Farmers' environmental characterization systems based on indigenous technical knowledge can be combined with currently used methods for analysing rice agroecosystems at the micro level. Such an approach can increase the effectiveness of on-farm prioritization by dealing directly with the multiple environments (on small scattered parcels) actually co-ordinated and managed by individual farmers.

Farmers mentally kept farm records by individual parcels, and the same was their actual unit of farm operation, implying that individual parcels should be the unit of data-gathering for farming research.

Farmers organized rice farming using diverse indigenous strategies inherited traditionally, but they also continuously modified and developed new farming techniques through their own kinds of experiments. Farmers undertook experiments for different reasons and had different types of results. The first rice experiment tested seedling-bed water-muddiness, seed source, water source, fertilizer, and, finally a new variety in order to find a spring rice cultivar and management practices that would not negatively affect the summer rice. Another rice experiment tested one cultivar in different environments to find it suited to a low, more water-logged environment. More important than the particular problems addressed, however, were some of the underlying features of these farmers' experiments.

- Farmers not only tried to solve existing problems (e.g. insufficient food supply), but also tried to create new alternatives (e.g. a spring rice), tested traditional practices (e.g. seed and fertilizer rates) and tested new questions (e.g. the effects of water source or temperature).
- Farmers experimented with both crop components and cropping patterns, viewed problems holistically (in terms of on-farm environmental diversity and the 'fit' of crops over time) and, therefore, automatically conducted 'systems research'.
- Farmers changed experimental designs, treatments, and variables in their repeating of experiments in succeeding seasons and years, based on their previous findings.

Both farmers and scientists are agricultural researchers. Achieving sustainable production under farmers' conditions would clearly be easier if partnerships were formed between experimenting farmers and experimental scientists. Attaining such partnerships, however, will require that scientists first value, and second understand, farmers' research capabilities. Rather than the learning of completely new practices, farmers may need more assistance in their research techniques at the 'adaptive research stage' of technology testing and diffusion.

Farmers in Nepal planted a wide range of TVs and MVs for a wide range of uses. Farmers sought, evaluated, and adopted or rejected varieties; and their selection criteria provide directions for varietal improvement research. Vigorous tillering with fewer unproductive tillers, fewer but deeper roots, shorter duration, good storage, high-quality straw, pest resistance, and tolerance to environmental fluctuations were farmers' rice-breeding targets. Farmers' indigenous tests of various aspects of rice varieties also reflected their varietal selection criteria, which could be categorized into environmental, technical, socio-economic criteria, and others. Consideration of such characteristics means that rice varietal improvement need not occur at the expense of existing useful varietal attributes.

2. Farmer-based experimentation with velvetbean: innovation within tradition

D. BUCKLES and H. PERALES[1]

Abstract

LOCAL KNOWLEDGE AND farmer perspectives were applied to the development of new cover-crop management strategies for maize-based systems in southern Veracruz, Mexico. Farmers' spontaneous innovation with velvetbean (*Mucuna pruriens*) was the basis for four cycles of farmer experimentation, resulting in a new management strategy acceptable to farmers (a mid-season velvetbean intercrop with the potential to ameliorate declining soil fertility, weed invasion, and drought stress). Farmer perspectives on maize/velvetbean associations are linked to the conceptual framework of shifting cultivation practised for centuries by the indigenous population.

Introduction

Recent appreciation of farmers' historical roles in developing and adapting traditional agricultural systems has focused attention on the potential of farmers to contribute to developing new strategies for sustainable agriculture. Scientists have tried to tap into the innate creativity of farmers and their intimate knowledge of their immediate environment through enhanced farmer participation in problem identification (Harrington *et al.*, 1992; Chambers, 1990; Fujisaka, 1989; Lightfoot *et al.*, 1988; Lightfoot, 1987) and the evaluation of agricultural technologies (Graf *et al.*, 1991; Ashby, 1990). Bentley (1994) points out, however, that examples of effective collaboration between farmers and scientists resulting in the development of new technology options are still rare. Ideas and methods for solving farmers' problems are typically sought in the realm of science, and subsequently introduced to farmers for adaptation to local conditions.

This paper examines the application of local knowledge and farmer perspectives to the development of new cover-crop management strategies for maize-based systems in the Sierra de Santa Marta in southern Veracruz, Mexico. Spontaneous farmer innovation with velvetbean (*Mucuna pruriens*), a vigorous, annual climbing legume, is described and new velvetbean management options developed through collaborative experiments are discussed. Farmer perspectives on maize–velvetbean associations are linked to the conceptual framework of shifting cultivation practised for centuries by the indigenous population, bringing to light the creative tension between tradition and innovation characteristic of local knowledge systems. The research indicates that local knowledge can be a fruitful point of departure for collaboration with researchers, and farmer participation an effective tool in developing new technology options acceptable to farmers. It also shows, however, that potential complementarities with science-based knowledge and experimental methods should not be overestimated. Insights gained through farmer participation in key research decisions are made at considerable cost to the reliability of agronomic data and should not be seen as a substitute for more conventional on-farm experimentation, but rather as an early phase of research on new technologies.

Farmer innovation with velvetbean

The Sierra de Santa Marta rises steeply from sea level at the Gulf of Mexico to more than 1700 m at its highest peak (Map 1). The soils of the Sierra (Andosols and Alfisols) are moderately fertile but highly susceptible to erosion. More than 3000 mm of rain fall annually on the northern, seaward slope of the range while on the southwestern slope annual rainfall averages some 1500 mm. The rains begin in late May or June, peak in October, and gradually drop off through January and February. A short, sharp dry season extends from April through May, interrupting most agricultural activities.

Climatic conditions facilitate two growing seasons per year, the *temporal* or

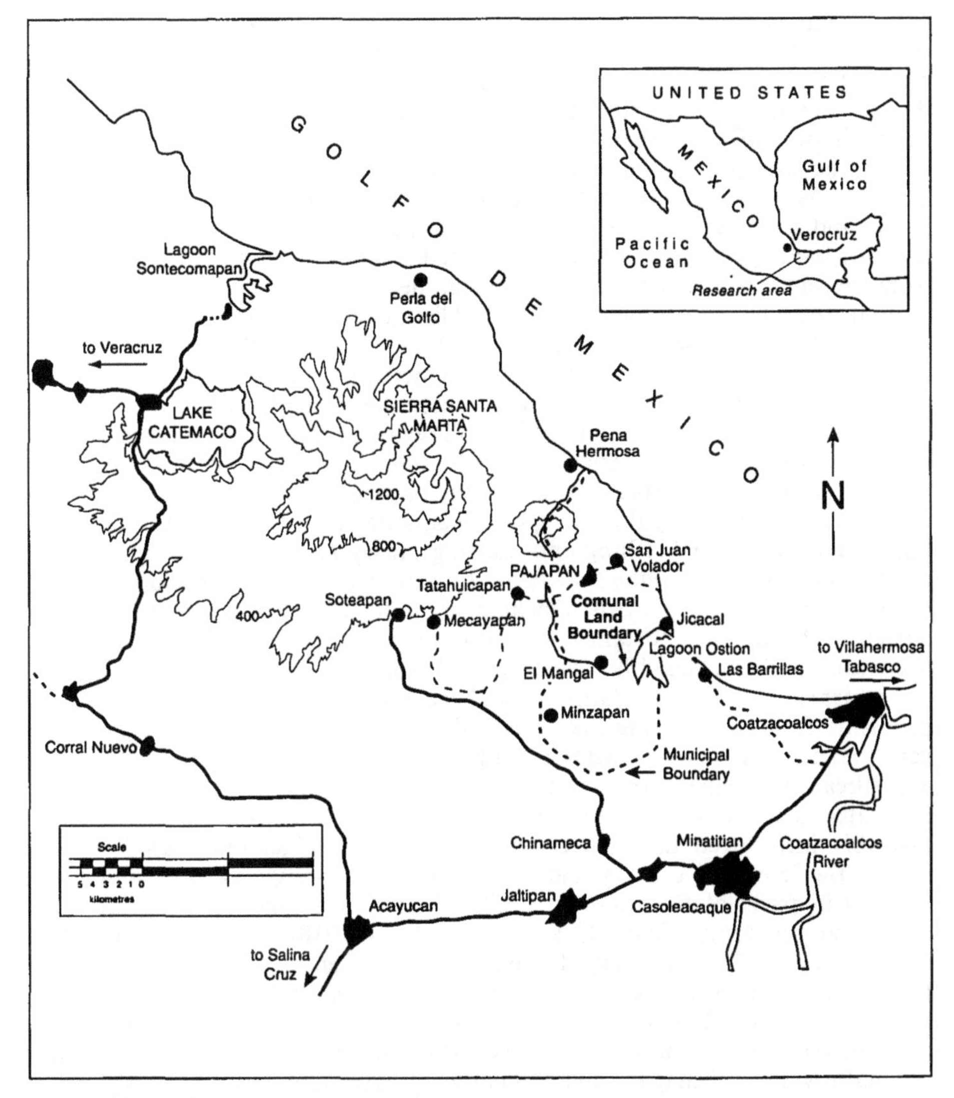

Map 1 Sierra Santa Marta research area

summer season and the *tapachole* or winter season. A wide range of annual and permanent crops can be grown in the Sierra, but maize and beans are the most important to the indigenous population. Nahua and Popoluca speakers have inhabited the Sierra since pre-Hispanic times, using techniques of shifting cultivation to produce basic foodstuffs. Fallow land, known as an *acaual,* is slashed and burned prior to the summer season in preparation for planting maize in June. Only local varieties are used. Crops are weeded manually or in combination with herbicides and doubled after reaching physiological maturity to facilitate drying in the field. Fertilizer use is rare, mainly due to cash constraints.

Periodic rainfall from November through February permits a winter maize crop to be grown in most of the Sierra. Winter maize is planted between the rows of doubled summer maize. Weed and crop residues from the summer season are not burned prior to planting winter maize, but rather are chopped and left on the field as mulch to conserve soil moisture. Because of the relatively dry winter conditions, maize plants do not need to be doubled prior to harvest and only one weeding is required, with the advantages of lower labour costs compared with the summer season. Nevertheless, maize can be flattened by strong winds during the early part of the season, and the risk of crop failure from drought is high during the later part of the season. Yields of winter maize average some 800 kg/ha, whereas summer maize yields average 1.4 kg/ha. However, intercrops such as beans and cassava, semi-permanent crops such as plantains, volunteer plants used for food (*quilties*), and tree species used for firewood may also be harvested from the same plot, mainly during the summer season.

While these basic techniques have varied little over centuries of use in the region (Stuart, 1978), the broader environment has. Since the 1950s, the expanding cattle industry and population growth have increased pressure on land resources, resulting in rapid deforestation, more intensive cropping patterns and land degradation. Although, traditionally, fallow periods of eight years or more were used to recover the agricultural potential of cultivated land, fallows have been reduced to two or three years in many parts of the Sierra. Fallow fields are no longer dominated by the tree species characteristic of the traditional fallow, but by grassy weeds. Slashing and burning grassy fallow cannot support crop production without drawing heavily on the limited resources of the soil and hefty investments in weed control. Declining soil fertility and weed invasion are currently the most important constraints on maize productivity in the region (Chevalier and Buckles, 1995; Pare *et al.*, 1993; Perales Rivera, 1992).

Farmers in the Sierra de Santa Marta have not been passive in the face of these problems. Interviews during 1991 detected a number of regional farmers growing velvetbean in their maize fields to improve soil fertility and eradicate weeds. Velvetbean is originally from India or China, where it was at one time widely cultivated as a green vegetable (Wilmot-Dear, 1987, 1984; Burkill, 1935). It was probably introduced into the Carribean by East Indian indentured workers during the late 19th century and later adapted by orange growers and maize farmers in parts of the southern United States as a soil-improving crop and cattle and pig forage (Scott, 1919; Tracy and Coe 1918). Transnational banana companies may have introduced velvetbean into Central America early in the twentieth century as a forage crop for mules (Buckles, 1995). It was subsequently adapted to maize-based farming systems among the Ketchi of Guatemala (Carter, 1969), the Chontales of Tabasco (Granados Alvarez, 1989), the Mames of south-western Chiapas, and Chinantecos (Arevalo Ramirez and Jimenes Osornio, 1988) and Mixes (Narvaez Carvajal and Paredes Hernandez, 1994) in the Isthmus area of north-western Oaxaca. It also found its way into the Sierra de Santa Marta in southern Veracruz, where it drew the attention of Nahua and Popoluca farmers.

Two distinct management practices were identified during informal field surveys. Farmers in San Pedro Soteapan, the cultural centre of the Sierra Popoluca, indicated that they had encountered velvetbean growing wild in their fields and noted its ability to smother weeds and improve maize yields. They collected seed and broadcast it over a larger area, giving rise to a practice known as 'making a fallow field' (Sp., *hacer acaual*). Farmers use velvetbean seed on a maize field they intend to fallow because of declining yields and weed invasion, typically at the end of the summer maize cycle. The crop develops for several months, sets seed, and naturally re-establishes itself during subsequent seasons. According to experienced farmers, weeds are eliminated by the aggressive velvetbean crop and soil fertility is regained: maize yields on land 'improved' with velvetbean for two years rival yields on land fallowed for five years with native trees and shrubs.

Fallow enhancement with velvetbean seems to be an endogenous innovation in direct response to declining land and labour productivity resulting from reduced fallow periods. Farmers can increase the frequency of cultivation without recourse to external inputs and yet maintain acceptable levels of production. The practice of 'making a fallow field' with velvetbean is recognized locally as an improvement brought to the field by the farmer, and even helps establish customary rights to particular parcels of land.

In 1991, an estimated three-quarters of the *ejidatarios*[2] of San Pedro Soteapan reported the presence of velvetbean in their fields (Perales Rivera, 1992). It had appeared spontaneously in the fields of half the growers, while the rest had originally planted the crop. Nevertheless, the management of velvetbean to improve soil fertility and control weeds is somewhat haphazard. Farmers do not plant velvetbean systematically on all fields abandoned to secondary vegetation or plant the crop every year, but rather rely on natural propagation. As a result, the velvetbean-stands observed in San Pedro Soteapan were uneven and scattered in small pockets within farms. Few farmers had velvetbean fields larger than 0.5 ha.

The casual qualities of improved fallow management using velvetbean observed in San Pedro Soteapan contrast sharply with a more systematic strategy noted among Nahua farmers in the neighbouring *ejido* of Mecayapan. Farmers reported that velvetbean is stick-planted at the end of the dry season in winter maize fields and allowed to develop as a sole crop throughout the summer season. The abundant velvetbean growth, known locally as a *picapica*, is slashed in November, and winter maize is stick-planted into the mat of decomposing leaves and vines, where it develops relatively free of competition from weeds. In keeping with traditional winter maize cropping practices, farmers do not burn velvetbean residues or incorporate them into the soil but leave them on the surface as mulch. Velvetbean seed that has matured in the field germinates during the winter, naturally reseeding the *picapica* without significant additional labour inputs. Some farmers reported more than 10 years of continuous cropping of winter maize following velvetbean.

Velvetbean summer rotations used by farmers in Mecayapan resemble strategies employed by farmers in Tabasco (Granados, 1989), Guatemala (Carter, 1969), and Honduras, suggesting that the practice may have been borrowed and adapted from other areas. Farmers from Mecayapan first planted velvetbean in the late 1940s in an unsettled area known as Pozo Blanco in the current *ejido* of Reforma Agraria[3]. The practice was later adapted to steeper hillsides when land reforms forced farmers to limit agricultural activities to their home village.

Farmers' use of velvetbean rotations with winter maize in Mecayapan seems to have been stimulated by significant land and labour productivity advantages compared with traditional winter maize cropping practices, rather than land degradation processes *per se.* Agronomic evaluation of the rotation elsewhere indicates that winter maize yields are higher and labour costs lower following velvetbean compared with traditional cropping practices, observations consistent with farmers' evaluations in Mecayapan (Triomphe, 1995; Flores, 1994). The velvetbean crop can produce more than 10 t/ha of dry matter during a summer season, resulting in a chemical fertilizer substitution rate in winter maize of approximately 150 kg N/ha. The conservation of soil moisture and the weed control effects of the velvetbean mulch also contribute significantly to improved maize yields.

An estimated 40 per cent of the farmers of Mecayapan cultivated velvetbean in a summer rotation with winter maize in 1991, in many cases growing all of their winter maize in this manner. Numerous fields ranging from less than 1 ha to over 4 ha were observed in a section of the *ejido* known as Cerro Tambor, where velvetbean use is currently concentrated. Velvetbean summer rotations with winter maize were also observed among a handful of farmers in the *ejidos* Mirador Saltillo, Tatahuicapan, Zapoapan, and Pilapillo, all Nahua communities.

Interviews in other communities in the Sierra, including the major villages in the municipalities of Soteapan, Mecayapan, and Pajapan, indicated that while many farmers in the region were familiar with the plant, very few cultivated it in their fields. In all, an estimated 150 farmers in the Sierra, concentrated in the *ejidos* San Pedro Soteapan and Mecayapan, used velvetbean in association with maize in 1991. While the plant and knowledge of its uses could be traced back several decades, use of the crop by most farmers interviewed dated from the early 1980s.

The pattern of farmer innovation noted above may partly explain the limited diffusion of velvetbean rotations and improved fallow in the region. Land degradation was not severe in most communities of the Sierra until the 1980s, when the agricultural frontier was exhausted by the expanding cattle industry and population growth. Before that time, most farmers could satisfy their maize needs using traditional techniques of shifting cultivation and had no reason to seek more intensive following practices. Similarly, velvetbean summer rotations with winter maize offer no opportunities in communities where erratic winter rainfall and strong winds severely limit winter maize production. Even in communities where some winter maize is grown, such as Soteapan, not all land is suitable for the crop, and many farmers consider the winter cycle too risky.

Lack of access to accurate information on the management and benefits of velvetbean may also have slowed diffusion. Velvetbean, known to experienced farmers as *picapica mansa* or 'tame picapica', is frequently confused by regional farmers with a species of velvetbean known as *picapica brava* or 'wild pica', which produces severe itching. Farmers indicate that they avoid use of this plant, considered a hazard in maize fields.

Changing land use patterns, particularly the expansion of pastures, may also have created barriers to adoption. Open grazing by cattle will eradicate velvetbean stands, and the use of fire to manage pastures and agricultural land threatens velvetbean stands and maize crops planted in dry velvetbean mulch.

But perhaps the most apparent barrier to diffusion of enhanced fallow and summer rotations with velvetbean was the opportunity cost of land dedicated to a soil-improving crop. Although velvetbean can be used to reduce fallow periods or produce winter maize, increasingly intensive cropping patterns make it costly for farmers to leave land fallow under velvetbean during the summer cycle, the most important widespread maize-cropping season. As noted above, rapid land use changes and population growth have forced many farmers to intensify land use patterns, despite declining productivity. Farmers who cannot afford to leave their land fallow during the summer cycle opt instead to fallow land during the winter cycle or increase their dependence upon external inputs. This problem became the focus of on-farm experiments with velvetbean.

On-farm experimentation with velvetbean

The local response to problems of soil-fertility decline and weed invasion in the Sierra de Santa Marta provided an opportunity to build on farmers' innovations, and the experience of a local farmer from Tatahuicapan indicated how this might be done. Experimenting with velvetbean, he allowed volunteer velvetbean plants to develop as an intercrop in his summer maize, pruning them back when they threatened to engulf the developing maize crop. Once the maize flowered, he stopped pruning the velvetbean and left it in the field to fallow the land during the winter season. The practice was repeated during the following summer cycle.

Consultation with agricultural scientists and relevant literature also suggested that velvetbean could be managed in association with summer maize. In on-farm trials managed by researchers at several sites in Central America, velvetbean was intercropped at the same time as summer maize to establish early ground-cover for controlling erosion, the opportunity to maximize the production of velvetbean competition with summer maize was considerable. Bunch (1990) reported on an independent experience with velvetbean intercrops in a subtropical region of central Honduras where only summer maize is grown. Farmers planted velvetbean simultaneously with maize and pruned the intercrop down to knee level when it threatened to compete with the maize. Some farmers incorporated the velvetbean residues by hand into the soil while others left the residues as a mulch on the soil surface. Positive effects of the practice included reduced labour requirements for weeding and improved maize yields in subsequent cycles.

These experiences suggested that a velvetbean/summer maize association might provide land-constrained farmers with the opportunity to improve their land without losing a summer season. It was not clear, however, how the crop should be managed or what benefits could be expected in the near and longer term. Consequently, on-farm trials were established to develop velvetbean associations with summer maize acceptable to farmers and to evaluate the potential impact of the technology. The approach taken by the authors was to stimulate further farmer experimentation with velvetbean by providing farmers with seed, new management options, and an experimental structure for evaluation in collaboration with researchers.

Materials and methods

General assemblies were organized through the local land authorities in three villages in spring, 1991. Farmers in two of the villages, Soteapan and Mecayapan, had previous experience with the crop; in the third village, Pajapan, the use of velvetbean was virtually unknown. The authors hoped that by including both farmers who were familiar and those who were unfamiliar with velvetbean, the farmers would be encouraged to participate in designing trials and provide varied perspectives on velvetbean's potential (Perales Rivera and Buckles, 1991).

During the meetings, attended by 15 to 40 farmers in each village, the velvetbean management practices observed in the region and potential associations with summer maize were described. Farmers were invited to select from the 'basket' of options for establishing trials on their farms alongside maize grown without velvetbean. Options included the local practice of a summer velvetbean rotation with winter maize and two new practices: velvetbean relayed into summer maize 50 days after maize planting and velvetbean planted simultaneously with summer maize. The researchers proposed that the velvetbean crop be planted between maize rows, using two or three seeds per hill and a planting distance of approximately 1 m between hills. Double cropping with winter maize and fallowing during the winter cycle were also proposed as options that farmers could try. Each option represented a distinct degree of intensification of cropping pattern and management, from rotations (the least intensive) to simultaneous intercrops with double cropping of maize (the most intensive).

For ease of farmer implementation and evaluation, large plots were proposed with only one treatment replication per location. A local land unit (*tarea*) measuring 25 m on each side (625 m^2) was used for each plot. Farmers wishing to attempt more than one option were encouraged to do so and were requested to accompany each option with its corresponding 'control' plot without velvetbean. The conditions required for controlled comparisons between plots (similarity of slope, soil conditions, and field history) were discussed. During the first cycle (summer 1991), 32 farmers established trials with velvetbean. All but eight of these farmers had no previous experience with velvetbean, although all had seen the plant in the wild or in other farmers' fields. Most farmers chose velvetbean associations with summer maize, a clear indication of a strong interest in the more intensive velvetbean management strategies. Only nine farmers selected the summer velvetbean rotation strategy, typically because they wanted to see if velvetbean could recover particularly degraded fields.

Initially only two cycles of trials were contemplated, but these were later extended to four cycles in response to farmer and researcher interests. Although the number of participating farmers declined[4] and attendance at group meetings was variable (at times, poor), a core group of 22 farmers interacted regularly with the authors during four cycles over 1991–93. The number of fields available for formal agronomic evaluation was much fewer.

Although self-selecting, the farmers participating in the trials were not untypical for their villages. All held rights to *ejido* land in parcels ranging from 4 to 12 ha, and most depended mainly on production from their *milpas* (maize fields) for their livelihood. Three farmers owned up to seven head of cattle, and several others frequently engaged in off-farm employment as day workers and petty traders. A wide range of age groups was represented. No women farmers participated, although one farmer's spouse occasionally attended group meetings in his place.

Researchers visited farmers' fields several times during each cycle to discuss the trial and collect data on field operations and crop development[5]. At the end of each cycle, yield was measured in standard-sized subplots for selected fields, and semistructured interviews were conducted with farmers in their fields and at home, both individually and in groups. Several group field visits were organized and notes were taken during numerous casual conversations concerning farmers' opinions, preferences, and ideas regarding the use of velvetbean. Farmers were frequently reminded that the researchers were uncertain about the potential of the technology and for that reason valued farmers' observations highly.

The information collected was shared with farmers and used to inform decisions regarding trial design and implementation. For example, during meetings at the end of the second cycle, farmers expressed an interest in comparing the effects of velvetbean and chemical fertilizers, resulting in the addition of a fertilizer treatment. The researchers suggested a herbicide treatment using Atrazine, a pre-emergent herbicide unknown to these farmers, to reduce early weed competition with maize and intercropped velvetbean. (Atrazine is a less toxic herbicide than Paraquat, which was already used by farmers). These treatments served as checks for the legume treatment and novel points of comparison for farmers.

To incorporate these modifications, the original plots were divided in two and the fertilizer treatment placed on half of the control plot and the herbicide treatment on half of the plot with velvetbean; these two were assigned at random within the main velvetbean or control treatment. Because Atrazine required a new method of herbicide application, the field assistants were trained to apply it to the corresponding plot at planting. They also supervised the application of inorganic fertilizer, buried at the base of the maize plant, at a total rate of 87–46–00 (the local practice), 41–46–00 applied at planting and 46–00–00 at 30 days. Farmers managed all other aspects of the experiment, including planting date for the velvetbean intercrop. While farmers had the last word, the researchers did not hesitate to express their opinions, request co-operation in establishing a relatively uniform trial across farms, and propose new ideas for experimentation.

The timing of velvetbean intercrops

Rejection of the early velvetbean intercrop was the first collective insight developed during the trial. Many farmers had initially resisted the idea of intercropping velvetbean at the same time as summer maize for fear that the intercrop would strangle the maize. The authors argued, however, that velvetbean could be pruned to avoid competition and that the velvetbean biomass accumulated during the summer season would probably have a noticeable positive effect on subsequent maize crops. To facilitate discussion of this option, an outing was organized to the field of the farmer in Tatahuicapan already managing velvetbean in close association with summer maize. This farmer pruned the velvetbean intercrop to avoid competition, much like farmers in central Honduras. While not all of the farmers were convinced by this visit, nine selected the treatment for trial. Six of these farmers eventually intercropped velvetbean within seven days after planting maize.

Researchers observed that velvetbean planted simultaneously with maize had grown considerably by the end of the summer cycle, far more than velvetbean intercropped 50 days or more after maize. However, yield data collected during the first cycle from four fields indicated that velvetbean sown simultaneously with maize had a negative (but not statistically significant) yield effect of approximately 300 kg/ha, possibly due to competition (Table 2.1). This finding is consistent with researcher-managed trials in Central America (Zea, 1991)).

Table 2.1. Yield effects (kg/ha) of velvetbean sown simultaneously with summer maize, Sierra de Santa Marta, Veracruz

Field	*Maize yield* With velvetbean	Without velvetbean	Difference
1	1340	2210	−870
2	580	710	−130
3	340	1030	−690
4	1600	1160	+440
Average	965	1277	−312

Note: The treatment effects of four harvested subplots were consistent within each location.

Only one farmer actually pruned the velvetbean intercrop sown at maize planting. He noted that two prunings were required to keep the crop under control, an investment he would be unwilling to make on his field as a whole. Based on his experience, an estimated five person-days/ha would be required to control the crop, a 6 per cent increase in the time normally invested in maize production. He and other farmers pointed out that constant attention would be required to know when to prune velvetbean to reduce competition with maize. The vigilance required to control velvetbean was at least as important a consideration as the time needed for pruning it. To the authors' surprise, the Tatahuicapan farmer who had managed velvetbean in close association with summer maize for years had also abandoned the practice, citing the high labour costs of controlling the crop.

Collaborating farmers were unwilling to invest labour in controlling early velvetbean intercrops, opting instead to avoid direct competition with maize's rough timing. Nevertheless, they were impressed by the large amount of velvetbean growth that could be attained during the cycle with an early intercrop. This experience led farmers to propose a midseason intercrop date for the second year of the trial, with a view to minimizing labour costs and competition while at the same time optimizing the production of velvetbean biomass. However, no uniform planting date was agreed upon. Farmers noted that on relatively fertile land velvetbean developed very quickly and could easily engulf the maize crop, even when velvetbean was planted 30 days or more after maize. On less fertile land, velvetbean developed more slowly and could be intercropped 30 days after maize without risk of competition. Spacing of the maize crop and timing of maize doubling, both factors that influence how much light penetrates the maize canopy, also appeared to influence velvetbean growth. For example, velvetbean planted late in one field had only two to four vines 30 to 50 cm long at the time of maize doubling. Forty days later the legume had completely covered the maize crop and, when slashed for winter maize, left a thick, evenly distributed mulch. Finally, cropping patterns also influenced farmers' decisions regarding the timing of the intercrop. When only summer maize was planned, farmers tended to intercrop velvetbean somewhat later (45 to 60 days after maize), whereas farmers planning on using the same field for winter maize tended toward earlier intercrops (30 to 45 days after maize). Thus, planting dates that minimize labour inputs and risk to maize and optimize the production of velvetbean biomass may range from 30 to 60 days after maize, depending upon a number of field and farming system factors

Table 2.2. Factors influencing the timing of velvetbean intercrops in summer maize

Factor	*Earlier establishment*	*Later establishment*
Competition		x
Fertile land		x
Maize densely planted	x	
Early maize doubling planned		x
Winter maize planted	x	

(Table 2.2). Costs of establishing the legume association within this range are estimated at approximately two person-days/ha.

Farmers' modifications to the timing of velvetbean/summer maize associations and explanations for their decisions suggest that immediate labour considerations may take priority over potential benefits provided by green manuring. While research with velvetbean indicates that early intercrops can produce considerable longer-term benefits (Zea, 1991), from the farmers' point of view the labour needed to control velvetbean is simply prohibitive. Thus cost–benefit analyses that treat the various inputs and outputs of a new technology equally may underestimate the importance placed on labour considerations by these farmers. Farmer trade-offs between short-term labour costs and longer-term benefits may be suboptimal from a strictly accounting point of view, even when appropriate discount rates are used.

Cropping patterns and residual effects of velvetbean associations

By the end of the trial period, two distinct velvetbean/summer maize associations emerged. Nine farmers intercropped velvetbean in summer maize as early as possible without risking competition with the maize and slashed the crop in preparation for winter maize. This strategy allowed for two maize crops annually, the most intensive of the velvetbean management practices. The remaining 13 farmers relayed velvetbean into summer maize somewhat later in the season (50 days or more after maize) and left the land fallow under velvetbean during the winter cycle. This strategy, while a less intensive use of land, takes advantage of September-January rainfall not utilized by the summer maize crop to grow a green manure crop. Velvetbean dies off in January or February due to physiological maturity and drought stress, leaving behind velvetbean seed and residues.

A critical question concerning the feasibility of these practices is the degree to which they help resolve important maize production problems, such as poor maize yields and weed invasion. Researchers' evaluation of maize yield and weed control effects of velvetbean associations was hampered, however, by problems with experimental design and farmer attrition. As noted earlier, large experimental plots were established with only one treatment replication per location. Initially, an incomplete, unbalanced block design was proposed, with individual farm trials treated as blocks. While this design may have yielded reliable agronomic data with the large number of trials initially established (32), fewer of the original fields were available for data collection as the trial progressed. Many farmers left the velvetbean plot fallow during the winter cycle, limiting the number of potential observations. Land tenure conflicts eliminated some trial sites. Some trials were harvested prematurely or damaged by accidental fires or bird attacks. In several cases, analysis of yield results was confounded because farmers refused to maintain the control plot free of velvetbean, preferring instead to extend the area under velvetbean. Furthermore, farmers' modifications of the trial design, especially burning of crop residues prior to planting summer maize, reduced the number of trials with comparable and complete data. Consequently, the experimental design was changed to treatments with and without velvetbean, requiring a simple paired 't' test for analysis. Criteria for determining the completeness of data included the uniformity of trial management and 'normal' development of velvetbean during the first cycle, relative to the rest of the fields. Nevertheless, the observations made by the researchers were too few to be statistically sound.

While the formal analysis of maize yield and weed control effects suffers because of the small sample and considerable variation between farmers, the data nevertheless illustrate broad tendencies in agronomic performance. First, it would appear that velvetbean growth in the Sierra de Santa Marta is subject to much variability, especially during the first year of establishment. During the first summer season velvetbean grew vigorously on only a few fields, leaving a dense, leafy mulch. More commonly, velvetbean grew well but did not completely cover the fields by the end of the season, leaving a thin mulch on the ground. In a few fields, velvetbean growth was poor; plants remained small and ground cover was limited[6]. During the second summer, however, velvetbean grew vigorously on most fields where it had performed fairly or only poorly the year before. Qualitative examination of soil conditions and laboratory analysis of soil samples in selected fields suggested that the variability and generally low level of fertility in these soils may have caused the variation observed in velvetbean development. While inconclusive, these observations suggest that two years of velvetbean intercropping may be required before the crop develops vigorously.

Second, the yield data suggest that the level of maize response to velvetbean associations is determined by the timing of subsequent crops. Yield gains during the 1993 winter cycle could be attributed to velvetbean associations with summer maize (Table 2.3). Although our sample for this cycle is small, velvetbean effects were strong and consistent; the plot with velvetbean yielded twice that of the control, although absolute yields were low due to a severe drought. Maize plants established in velvetbean mulch during the winter cycle were taller, with stronger stems and darker green leaves than maize in the control plot without velvetbean mulch. Moreover, in the velvetbean plot the number of plants per hill surviving the season's drought was twice that of the control plot (2.06 vs. 0.97; P 0.05), an indication that the velvetbean mulch helped conserve enough soil moisture for some plants to develop.

Table 2.3. Maize yields at three experimental sites during the winter cycle of 1993, Sierra de Santa Marta, Veracruz, Mexico

	Maize yield (kg/ha)	
Site	*With velvetbean mulch*	*Without velvetbean mulch*
1	602.5	223.2
2	589.1	292.9
3	305.6	25.2

Table 2.4. Average yields (kg/ha) during the four cycles of farmer-based experimentation with velvetbean (VB) in the Sierra de Santa Marta, Veracruz, Mexico

Cycle	*n*	*With VB mean (SE)*		*Control mean (SE)*		*P*
1 (wet)	13	1,623	(215)	1,653	(210)	>0.50
2 (dry)	3	650	(117)	568	(44)	>0.50
3 (wet)	8	1,245	(168)	1,345	(233)	0.30
4 (dry)	3	499	(96)	180	(80)	0.01

Note: P is the significance level for a two-tailed paired 't' test.

In contrast to the positive effects noted during the winter cycle, no significant differences between the control and velvetbean treatments were found during the summer cycle (Table 2.4). High variability between farmers may partly explain the non-significance of these results; differences between the control and velvetbean treatment ranged from $^{-}$420 to $^{+}$760 kg/ha during the first summer season and remained high during the second summer season. Despite high variability, however, a statistically significant response to the application of fertilizer was encountered; maize yields were more than 700 kg/ha higher on plots receiving fertilizer compared with other treatments (P probability <0.0001). This response indicates that even though the variation between farmers was high throughout the experiment, a clear treatment effect could be determined under the experimental conditions. Thus the absence of a significant effect was more likely due to the modest development of velvetbean on many fields and very high rates of mineralization of velvetbean residues during the winter cycle. Even on fields where velvetbean growth had been good in the previous year, very little velvetbean cover remained on the ground at the beginning of the following summer cycle. While inconclusive, these observations suggest that the benefits of velvetbean/summer maize associations may not carry over into the following summer cycle, or are too weak to measure under farmer-based experimental conditions.

Velvetbean effects on weed populations followed a similar pattern. Weed measurements were taken 20 to 25 days after maize sowing in 10 randomly chosen, 1m squares delimited by four hills. In fields with vigorous velvetbean growth in the summer cycle, weed suppression was apparent during the following winter cycle; maize with velvetbean had less weed cover (20 per cent or less) compared with the control field without velvetbean (60 per cent weed cover). This sizable effect possibly resulted from weed suppression caused by velvetbean competition with weeds during the growing season. The effect of the velvetbean intercrop was apparent during the following summer season. Again, the limited development of the velvetbean crop during the first cycle and the small amount of residue left on the field at the beginning of the following summer cycle may have accounted for the absence of observable weed control effects in summer maize.

These observations raised concern over the potential impact of velvetbean intercrops on summer maize. The agronomic data and researchers' assessments, although not conclusive, suggested that the benefits of velvetbean intercrops in summer maize were limited to cropping patterns involving winter maize, at least during the first year or two after establishment. Summer maize did not seem to benefit immediately from management of velvetbean in the previous summer season. This seemed to be true both in fields where velvetbean was left to develop as a winter fallow and in fields where winter maize was planted. While conventional agronomic trials with adequate repetitions and control of experimental error may have detected measurable differences in summer maize attributable to velvetbean associations, these effects would probably have been limited and would certainly have been continued to demonstrate great variability.

Farmers' objectives and evaluation criteria

Farmers' evaluation of velvetbean associations with summer maize, while generally consistent with researchers' observations, was much more favourable. Farmers confirmed that a velvetbean association in summer maize, followed immediately by winter maize, resulted in a notable yield and weed-control benefit during the winter cycle. These effects were attributed by farmers to improved soil fertility, soil moisture conservation, and weed suppression by the velvetbean crop, a perspective compatible with researchers' observations. Farmers also claimed, however, that velvetbean planted in summer maize improved field conditions for subsequent summer maize crops – effects not captured in agronomic data collected by the researchers.

Differences between science-based and farmer evaluation criteria may partly explain discrepancies between researchers' and farmers' observations. Farmers' evaluation of the trials was more comprehensive and holistic than the partial, fragmented evaluation realized by the researchers, a perspective on technology now generally appreciated by researchers (Bentley, 1994; Byerlee, 1993; Ashby, 1990; Norman *et al.*, 1989). While the researchers focused on short-term yield and weed control effects, farmers were equally concerned with the multiple and incremental effects of velvetbean associations on field conditions. During interviews in farmers' fields at the end of each cycle, farmers reported a wide range of positive effects of velvetbean associations with summer maize, including improved soil fertility, reduced weed populations, better soil structure ('softer soil'), less soil erosion, better moisture conservation during the winter season, and reduced damage to maize from soil pests. When ranked in order of importance by farmers, these factors highlight the importance of soil fertility benefits and labour savings among farmers' priorities (Table 2.5). This perspective reflects the multiple objectives of farmers and their perception that the technology has the potential to respond to a number of production constraints simultaneously.

During the trial, farmers managed velvetbean in different ways to respond to different problems, sometimes even within the same field. Farmers concerned

Table 2.5. Benefits of velvetbean associations with summer maize identified by farmers (no. of farmers)

Benefits	*First selection*	*Second selection*
Improved soil fertility (*abono*)	12	5
Weed control (*aplasta malezas*)	4	12
Moisture conservation (*conserva humedad*)	2	2
Erosion control (*no se lava la tierra*)	1	0
No response	3	3
Total	22	22

Note: Farmers ranked velvetbean benefits using cards depicting the characteristics.

Table 2.6. Main reason given for selecting a velvetbean management option, first season (no. of farmers)

Main reason	*Option selected* Summer rotation	*Summer intercrop*
Eradicate weeds	6	3
Improve soil fertility	1	14
No reason given	1	3
Total	8	20

Note: Six farmers selected more than one option.

primarily with eradicating persistent weeds from their fields typically opted for velvetbean summer rotations, while farmers concerned mainly about declining soil productivity intercropped velvetbean in summer maize (Table 2.6). Six farmers diversified velvetbean management strategies within their fields, planting summer rotations or improved fallow in areas invaded by weeds and intercrops in maize fields. Several farmers managed parts of their fields with velvetbean planted 50 days or so after summer maize to improve soil conditions and control weeds during the winter cycle, while other parts of their fields were managed with a midseason velvetbean intercrop followed immediately by winter maize.

Farmer adaptation of velvetbean management strategies to particular field conditions and production constraints suggests that one strength of the technology is its flexibility in the face of diverse production environments. Velvetbean can be 'applied' to fields in different ways as needed, akin to a 'component' technology such as fertilizer or herbicides. An implication for research is that adoption of legume associations may be favoured by the development of a wide range of flexible management options rather than the refinement of a single, 'ideal' practice.

During the evaluations, farmers also expressed concerns about possible problems arising from the use of velvetbean in their fields. Many indicated that rats could become a problem as the pest is attracted to areas with protective ground cover. Some noted that rats had climbed up the vines of the velvetbean intercrop to feed on ripening maize plants. A few farmers felt that the amount of time required to harvest summer maize increased owing to the abundant growth of velvetbean vines covering the doubled maize. This was particularly problematic in cropping patterns involving winter maize, because land preparations were required long before the velvetbean crop had stopped growing.

Perhaps the most important limitation on velvetbean associations with summer maize noted by farmers was incompatibility with traditional intercropping and following practices. Velvetbean in summer maize competes directly with volunteer plants used for food (*quilites*), intercrops such as beans and cassava, semi-permanent crops such as plantains, and tree species used for firewood. While velvetbean associations provide a means of intensifying maize production, farmers noted that the strategy is appropriate only for parts of the field sole-cropped with maize.

Despite these constraints, virtually all farmers participating in the trial continued to plant velvetbean in summer maize fields independently of the researchers, and more than half of the farmers increased the area under the technology from one year to the next – compelling evidence of farmers' interest in the practice. Farmers noted that direct costs of the technology were limited to collecting seed and planting it. They also emphasized the cumulative benefits of various small, short-term changes in field conditions and expectations that longer-term benefits would follow. This perspective suggests that farmers are willing to invest in longer-term land rehabilitation so long as some short-term benefits are present, direct costs are low, and positive tendencies can be perceived. The possible role of normative criteria in farmers' evaluation of velvetbean associations is examined below in the light of the regional indigenous knowledge system.

Indigenous knowledge and technical change

Farmers' evaluation of velvetbean management strategies may draw heavily on prior knowledge of the beneficial effects of fallowing in shifting cultivation systems. Farmers in the region maintain that fallowing allows the land to 'rest' after continuous cropping[7]. Resting the land generates a large amount of 'litter' (Sp., *basura*; Nahua, *tasol*) that 'becomes soil' and 'penetrates the earth' by rotting and keeping the soil wet, cold, or fresh (Sp., *fresca*; Nahua, *cece*). Fallowing is aided by 'cold' plant species (e.g., *jonote*) that 'give life to maize', 'soften the soil', and 'easily release the juices of the earth'. By contrast, 'hot' plants such as grasses (Sp., *zacates*) 'dry out' and 'harden' the soil. Fallow fields are called a 'maize house' in Popoluca (*poc tui*), a clear reference to the connection between fallowing and future maize harvests. A similar association is made by Nahua farmers (Stuart, 1978).

Farmers perceive the recovery of agricultural potential through fallowing as a process of 'healing' obtained through a balanced combination of the cold and the hot: a maize plant 'does not want sun or heat only, nor does it want water or cold only' (Chevalier and Buckles, 1995). From an indigenous perspective, the 'cold' conditions prevailing during the fallow period must be moderated by burning the trees and other shrubs slashed in preparation for planting summer maize. Management of these contrasts applies to annual cropping practices as well; while farmers know that mulching with crop residues during the winter cycle will enhance the 'coldness' and 'wetness' of the soil during the hot, dry winter season, they also count on the spring burning of plant residues to avoid excessive 'wetness' during the summer cycle.

The conceptual framework of land management through fallowing informs farmers' adaptations of velvetbean. 'Improving the fallow' with velvetbean and practising velvetbean summer rotations mimic the functions of fallowing. Farmers note that a thick canopy of velvetbean growth shades out grassy weeds while the decomposing leaves shed continuously throughout the growing season 'become soil' (Sp., *se vuelve tierra*) and 'fertilize' (Sp., *abona*) the land. According to farmers, velvetbean is a 'cool' plant that 'freshens the soil' and 'burns' weeds. Farmers described the impact of velvetbean associations with summer maize in these same terms.

Conceptual parallels between traditional fallowing practices and green manuring practices may also have facilitated farmer experimentation with the technology. During the trials, more than half of the farmers (15) experimented with velvetbean outside of the experimental plots in ways not considered by the researchers. Several farmers tried managing velvetbean as a living mulch by heavily pruning (but not killing) the crop prior to planting winter maize. Various velvetbean plant densities and spatial arrangements were also attempted.

Velvetbean was planted by experimenting farmers as an intercrop in winter maize, a relay crop with cassava, and a cover crop in mango orchards. If farmers' potential for adapting technology to specific situations is to be utilized fully, researchers should not underestimate the importance of farmers' understanding of the underlying logic of new technology.

Although indigenous knowledge may facilitate farmer innovation, some features of that knowledge may also impede comprehensive processes of technical change. As pointed out by Bentley (1989), farmers know more about some things than others and may have entirely erroneous notions about some issues. Dramatic environmental change may also undermine the basis of some local knowledge, and outstrip the capacity of local systems to develop new knowledge.

In the Sierra de Santa Marta, as in much of the humid tropics of Mesoamerica, the conceptual framework for shifting cultivation also calls for the use of fire during land preparation for the summer season. According to farmers, burning clears the land for planting, destroys weed seeds, reduces the incidence of maize diseases (black spot; *chahuiste*) and pests such as rats, and converts vegetation into nutrient-rich ash available to subsequent crops. While this strategy is well adapted to the management of mature fallows, it is much less compelling in systems where fallow periods are too short for significant regrowth. As indicated above, fallow fields throughout the Sierra are no longer composed of large trees, vines, and herbaceous growth but rather grasses and small shrubs that present few obstacle to planting. Grassy fallows do not burn hot enough to destroy weed seeds or the roots of many plants. Furthermore, very little ash is left on the field to nourish maize plants after annual weed and crop residues are burned. In short, under prevailing fallow practices, burning provides few benefits and imposes new costs, such as the loss of soil organic matter and soil erosion.

Farmers in the Sierra de Santa Marta are aware of the problems caused by continuous burning. Many note, for example, that excessive burning promotes the development of 'hot' plants such as grasses that compete with maize. Farmers' practices, however, have not changed; burning crop residues prior to planting summer maize is still the rule. Most farmers participating in the velvetbean trial continued to burn crop residues outside of the experimental plots, despite urging from researchers to the contrary. It seemed that the conceptual framework of shifting cultivation, with its emphasis on burning residues, was creating barriers to full realization of the potential benefits of legume associations.

To confront this problem, the authors examined farmers' reasons for burning crop residues and proposed alternative means of realizing some of their objectives without burning. For example, researchers proposed the experiment with Atrazine, which controlled weeds effectively in unburned plots. Several farmers in the experimental group began using the herbicide on their own fields when preparing land for summer maize, prompting them to conserve the crop residues. From their perspective, the herbicide had 'burned' the weeds and could be expected to foster the conditions needed for a successful summer maize crop.

The experiments with velvetbean also increased farmers' appreciation of the value of crop residues, stimulating several experiments with alternatives to burning. Two farmers attempted controlled burns (*media quemadas*) on days following heavy rains to reduce the bulk of surface residues and perceived risks of maize diseases and pests. Several others tried slashing all regrowth on their fields a number of months before planting summer maize to facilitate complete drying down of the residues, a process likened to burning. This strategy called for an additional weed-control operation immediately before planting summer maize.

These incipient attempts to resolve problems with conserving crop residues are more likely to receive serious consideration from farmers than simple prohibitions against burning. The research strategy suggested by this experience is to confront directly the problems traditional practices are meant to resolve and to link new skills to old ones, thereby helping farmers move beyond the boundaries created by their current knowledge system. Farmers constantly engage in a process of resolving tensions between tradition and the need for innovation. Researchers, through a synthesis of scientific and local knowledge, can accelerate this process.

Conclusions

This chapter has provided further evidence of farmers' continuing contribution to the development and adaptation of new technology. Clearly, farmers, farming systems, and farmer knowledge systems are not static. Farmers actively seek solutions to agricultural problems – experiments that can provide a point of departure for collaboration between farmers and researchers.

The experience in Santa Marta suggests that collaborative research can contribute significantly to technology development. Four cycles of farmer-based experiments led to relatively quick progress in developing a new management strategy acceptable to farmers. Farmers rejected an early intercropping strategy tried elsewhere in favour of a midseason intercrop, a low-cost strategy with the potential to ameliorate problems of declining soil fertility, weed invasion, and drought stress. This strategy was not previously available to these farmers.

Farmer participation in the design, implementation, and evaluation of on-farm experiments with velvetbean also identified farmers' management criteria that were relevant to future research with legume associations. First, labour considerations appear to take priority over potential benefits provided by legume associations, a perspective that may modify the weight given to labour costs in evaluating the costs and benefits associated with new management options. Second, farmers perceive legume associations as a multipurpose, component technology that can be used flexibly to respond to several production constraints simultaneously. The development of a wide range of management options with various impacts may be more likely to meet the diverse and multiple needs of farmers than refinement of one 'ideal' practice. Third, farmers may be willing to invest in legume associations with longer-term benefits so long as direct costs are low, some short-term benefits are realized, and longer-term benefits are perceived. Farmers' evaluation of legume associations may be influenced by normative criteria regarding expected benefits and potential costs, perceptions informed by the local knowledge system. Farmers' perspectives on velvetbean associations rely partly on the conceptual framework of shifting cultivation practised for centuries by the indigenous population. Research strategies that build on farmer knowledge and target limitations in that knowledge may accelerate the development of technologies acceptable to farmers and enhance farmers' input into technology adaptation.

While farmer-based experimentation with velvetbean was productive, the insights gained through farmer participation in key research decisions were made at considerable cost to the reliability of the agronomic data. The variability and seasonality of residual yield and weed effects of velvetbean was apparent but the trials were not up to the task of reliably measuring the magnitude of these effects. Trial design, farmer attrition, and across-farm variation in both experimental and non-experimental variables limited the experimental data available for valid treatment comparisons. A methodological implication of this outcome is that farmer-based experimentation should not be seen as a substitute for conventional on-farm trials but rather as an early phase of research on new technologies. Farmer participation can generate new ideas for experimentation and help establish the range of alternative strategies with the highest potential for adoption, thereby improving the efficiency of the technology generation process. Once these strategies are identified, researcher-managed trials can focus on key issues requiring quantification. Thus, the basis for effective collaboration is the recognition that both farmers and researchers bring unique skills to the task of developing new strategies for sustainable agriculture.

3. Side-stepped by the Green Revolution: farmers' traditional rice cultivars in the uplands and rainfed lowlands

S FUJISAKA

Abstract

SINCE THE mid-1960s, rice farmers in the irrigated areas of Asia have rapidly adopted 'Green Revolution' rices because of their responsiveness to nitrogen fertilizer and their higher yields, shorter crop duration, and shorter stature. Such cultivars were well suited to systems with good water control and moderate to high management inputs. Although modern rice cultivars have been adopted in less favourable environments, farmers also continue to rely on their traditional cultivars in the uplands and rainfed lowlands. Rice breeding strategies are now being developed that are more tailored to such unfavourable rice environments. Farmers' criteria for selecting or rejecting different rices in the unfavourable regions constitute a valuable resource for programmes interested in improving the productivity of such bypassed areas.

Introduction

Reported mean yields of upland (1.1 t/ha) and rainfed lowland (2.1 t/ha) rice are low compared with those of irrigated rice (4.7 t/ha) in 37 major rice-producing countries that are less developed (IRRI, 1988). Differences are largely due to inherent differences in rice-growing environments, but also because few improved varieties have been developed specifically for the uplands or rainfed lowlands. Rice breeding by the International Rice Research Institute (IRRI) and by national agricultural research programmes has in the past focused attention, reasonably enough, on irrigated environments in order to improve global food security quickly and substantially. They have only recently been able to turn to the less favourable, riskier rice agroecosystems.

Breeding strategies to improve irrigated rice emphasized the development of semi-dwarf (mainly Indica) rices of high yield potential, shortened growth duration (to about 100 days), and increased pest resistance or tolerance (IRRI, 1989). Much of the research was conducted on-station where conditions matched those of the more favourable irrigated areas. In many areas the result was that one (or at most a few) improved, high yielding, 'modern' variety (MV) replaced several traditional cultivars.

A similar 'Green Revolution' has not taken place in the uplands or rainfed lowlands. At IRRI, the strategy for improving rice in such environments was initially similar to that used for irrigated areas – a search for higher-yielding, semi-dwarf Indica rices, albeit cultivars suited to less favourable conditions. Fortunately, recognition that breeding strategies used for irrigated rice are inappropriate for the less favourable environments has now gained widespread acceptance.

Within the last 15 years, IRRI research to improve upland rices for Southeast Asia has been based on the Japonica rices grown by 85 to 95 per cent of upland rice farmers in the region. Japonica rices are characterized by definite tillering, long panicles of high grain count, and thick culms. Indica rices are characterized by indefinite tillering, short to medium panicles, and smaller diameter culms. Research is now ideally two-staged. First, IRRI provides advanced breeding lines that are adapted to poor or medium-poor soil, are blast resistant, and are to some extent drought resistant. National programmes then develop – through both on-station and on-farm testing – and then release varieties suited to local conditions and consumer preferences (M. Arraudeau, 1994, personal communication).

Similarly, for the rainfed lowlands, plant breeders now look for submergence and drought tolerance, a range of maturities, and degrees of photo-period sensitivity. Major efforts to improve population are on-going in South and Southeast Asia, with a strategy of building on traditional Aus (neither Indica nor Japonica)-type rices (R. Zeigler, 1994, personal communication).

Some research has been working to develop improved methods in the search for rices suited to highly risk-prone and diverse environments. Such work has demonstrated the importance of combining characteristics of farmers' traditional cultivars with those of advanced breeders' lines (Maurya *et al.*, 1988). It also shows the importance of testing materials in the poorer (i.e., farmers') environments rather than under the usually more favourable on-station conditions (Maurya *et al.*, 1988; Simmonds, 1991), and the strong willingness and sound decision-making of farmers participating in rice testing and selection (Richards, 1986; Haugerud and Collinson, 1990; Chaudhary and Fujisaka, 1992; Simmonds and Talbot, 1992).

This chapter examines farmers' choice, use, and evaluations of rice cultivars in the Philippines: of upland rices in Claveria and Bukidnon in northern Mindanao, and of rainfed lowland rices in Solana and Tarlac, northern Luzon. Yields are variable and often low. But farmers' choices of cultivars, evaluations of positive and negative characteristics, reasons for discontinuing cultivars in the past, and accounts of ideal rice characteristics provide a valuable set of resources for rice breeding programmes. This is especially so in terms of the practical knowledge that has allowed farmers successfully to select locally suited and adapted rices over many, many crop seasons.

Methods

Sixty-seven upland rice farmers were randomly selected and interviewed in several municipalities (Kalilangan, Damulog, Quezon, and Kibawe) of Bukidnon Province in the southern island of Mindanao, Philippines, and 37 upland rice farmers in the municipality of Claveria in the neighbouring province of Misamis Oriental. Then rice farmers in the rainfed lowlands of the northern Philippines and on the island of Luzon were randomly selected and interviewed, 32 in Solana, Tuguegarao Province, and 33 in several communities in Tarlac Province.

Each farmer was asked to identify rice cultivars he or she was growing in the current wet season, the area planted to each, good and bad characteristics associated with each, cultivars planted in the past but discontinued (with reasons for discontinuation), and traits desired in an ideal upland rice cultivar. In some cases, farmers described cultivars in contradictory terms. For example, some Bukidnon upland rice farmers liked the taste and aroma of the cultivar, Speaker; others so disliked the flavour that they would not eat the same rice. Respondents were also encouraged to discuss openly their rice cultivars, the reasons for their choices, and the ways they obtained or selected rice cultivars.

Data were described in simple tabulations. Yield data from Claveria were obtained from crop-cut samples. Yield data in the other cases were based on farmer recall.

Results

Upland rice

All the upland rice farmers in Bukidnon planted only traditional cultivars. Of 18 cultivars planted, one (Dinorado) dominated in terms of percentage of farmers planting (57 per cent) and area planted (a mean 0.86 ha planted per farmer growing Dinorado). Farmers characterized Dinorado as having good eating quality, high market price, pest and disease resistance, high tillering, high yield, high milling recovery, low shattering, lodging resistance, and some drought tolerance. Negative characteristics cited for Dinorado were lodging, pest and disease problems, susceptibility to drought, high losses to birds, and shattering (Table 3.1).

Fewer Bukidnon farmers planted Gakit, Azucena, Speaker, Pilit, and other traditional cultivars. These farmers sought similar positive characteristics (i.e., taste and price) and wanted to avoid the same negative characteristics (i.e., lodging, susceptibility to pests and diseases, shattering, and late maturation). Bukidnon farmers not growing Dinorado (43 per cent) included several with lands in higher, cooler areas. These farmers planted Gakit instead; 29 per cent of those planting Gakit described it as 'suited to the local climate'.

Upland rice has been planted in Bukidnon for at least 30 years. Over that period many respondents discontinued planting one or more traditional cultivars, including Azucena, Dinorado, Hinomay, Kurikit, and Lubang. Reasons for dropping cultivars included the expected ones of low yield, 'not adapted to poor(er) soils', drought susceptibility, shattering, and lodging. Other often-mentioned, but perhaps less-expected, reasons included cooked leftovers becoming hard, spiny awns making threshing (especially by foot) difficult, loss of taste in storage over time, weevil damage in storage because of thin husks, many off-types, greater susceptibility to insect pests caused by aroma at the vegetative stage, and insufficient stickiness in sticky rices (Table 3.2).

In open-ended discussions, several Bukidnon farmers reported discontinuing traditional cultivars such as Palawan, Hinomay, Ginatus, and Pamintana when they (several years before) adopted the now widely planted Dinorado, which had similar positive characteristics to the replaced cultivars but fewer of the undesirable characteristics.

Upland rice farmers in Claveria planted mainly the improved UPLRi5 (51 per cent of respondents; UPLRi5 is an Indica developed by the University of the Philippines at Los Baños) on a mean 0.84 ha, and the traditional Speaker (54 per cent) on a mean of 0.65 ha. Farmers favoured these rices for good taste, and high yield, tillering, and milling recovery. UPLRi5 was valued because its erect flag leaves reduce losses to birds, its shorter stature reduces lodging, it has some drought resistance, and is responsive to fertilizer. Conversely, UPLRi5 is of long duration, is susceptible to diseases, and its short stature leads to high losses to chickens and low weed competitiveness. Farmers liked Speaker because of pest and disease resistance, tolerance of acid soils, plant architecture that suppresses weeds, and high market price. Speaker's unfavourable traits included lodging and long duration (Table 3.3). Some respondents liked and others disliked the taste and strong aroma of Speaker when cooked.

Table 3.1. Farmers' (n = 67) upland rice cultivars and evaluations, several communities, Bukidnon, 1992 wet season

Characteristics	*Rice cultivars*											
	Dinorado	*Gakit*	*Azucena*	*Speaker*	*Pilit*	*Lukosomama*	*Panabang*	*Malam*	*Intramis*	*Binulawan*	*Kapakaw*	*Others†*
Farmers growing it (%)	57	21	13	11	10	6	4	4	4	3	3	15
Mean area grown (ha)	0.86	0.43	0.43	0.63	0.31	0.75	0.38	1.50	1.00	0.38	100	0.71
Positive characteristics:												
good taste/eating quality	89	64	67	75	29	100	100	67	67	100	100	60
taste does not deteriorate	8	0	0	0	0	25	0	0	33	0	0	0
left-overs remain soft	0	0	0	13	0	0	0	0	0	0	50	10
speciality food	0	0	0	0	29	0	0	0	0	0	0	0
good grain colour	6	7	11	0	14	0	0	33	33	50	0	10
high price if sold	63	21	22	13	71	0	0	33	0	50	0	20
easy to thresh	3	7	0	0	14	25	0	0	0	0	0	0
pest and disease resistant	78	29	33	38	71	25	100	67	33	100	0	50
high tillering	59	14	33	25	14	25	50	67	0	50	50	30
high yield	56	21	17	63	0	25	50	67	0	0	0	40
high milling recovery	53	43	33	50	29	0	0	67	33	100	0	40
does not shatter	41	29	44	13	14	50	0	67	0	100	100	20
does not lodge	37	64	56	38	29	50	100	67	0	100	100	70
resists short drought	24	21	22	0	43	0	50	33	33	50	0	10
responsive to fertilizer	21	14	11	13	0	0	50	0	0	0	0	10
tolerates weeds	13	14	11	0	14	0	0	33	0	0	0	10
thrives on poor soils	13	21	11	0	0	0	0	33	0	50	0	10
medium duration	11	7	0	38	0	0	100	0	0	50	100	10
heavy grain	9	21	11	13	0	0	0	0	0	0	0	20
good panicle exertion	6	7	22	25	14	0	0	0	0	0	0	10
not attacked by birds	3	7	11	13	14	0	50	0	0	0	0	10
Negative characteristics:												
None	29	29	33	25	43	0	0	33	67	50	100	20
taste deteriorates w/storage	3	14	0	0	0	0	0	0	0	0	0	0
hard to thresh	3	0	22	0	0	0	0	0	0	0	0	0
itchy to harvest	0	14	0	0	0	0	0	0	0	0	0	0
cooked left-overs harden	0	14	0	0	0	0	0	0	0	0	0	0

Table 3.1. Continued.

Characteristics	*Rice cultivars*											
	Dinorado	*Gakit*	*Azucena*	*Speaker*	*Pilit*	*Lukosomama*	*Panabang*	*Malam*	*Intramis*	*Binulawan*	*Kapakaw*	*Others†*
lodges	58	7	22	38	29	25	0	0	0	0	0	10
pests and disease problems	24	14	0	0	0	50	33	33	0	0	0	0
drought susceptible	21	0	0	13	0	0	33	0	0	0	0	10
bird susceptible	18	7	11	13	14	0	0	0	0	0	0	20
shatters	18	7	22	50	0	25	33	33	0	0	0	20
not good on poor soils	8	0	11	13	0	0	0	0	0	0	0	30
late maturing	5	14	0	13	14	0	0	0	0	0	50	0
lower tillering	5	0	0	0	0	25	33	0	0	0	0	0
small panicles	3	7	0	0	0	0	0	0	0	0	0	0
poor milling recovery	0	7	0	0	14	0	0	0	0	0	0	0
cross pollinates	0	0	0	0	14	0	0	0	0	0	0	0
poor stand	0	0	0	0	0	25	0	0	0	0	0	0
low grain weight	0	0	0	0	0	25	0	0	0	0	0	0
not competitive to weeds	8	0	0	0	0	0	0	0	0	0	0	0

† Lubang, Milagrosa, Palawan, Bulahan, Guyod, and 7 tonner were each grown by only one farmer.

Table 3.2. Upland rice cultivars no longer planted by farmers and reasons given, Bukidon, 1992 wet season

Cultivar	*Reason*
Azucena	Drought intolerant; long maturation; spiny awn makes handling and threshing difficult; bird damage because of early panicle exertion.
Basura	Weevil damage in stored grain caused by thin carp.
Binuhangin	Not adapted to poorer soils.
Binolawan	Cooked left-overs are hard; taste deteriorates with longer storage.
Dahili	Cooked left-overs are hard; aroma is lost with longer storage; shatters; lodges.
Dinorado	Not adapted to poor soils; high loss to birds because of small grain; disease, stem borer, and rice bug problems; lodges; not suited to cooler climate; taste deteriorates with longer storage.
Gakit	Shatters; cooked left-overs hard; poor taste; bad aroma when cooked; not adapted to poor soils; many off types.
Handurawan	Not adapted to poor soils.
Hinogayan	Low yield; poor taste.
Hinomay	Not adapted to poorer soils.
Inampik	Poor taste; not adapted to poor soils; hard left-overs; painful to thresh (i.e., by foot) because of sharp awns; unfilled grain in cooler areas.
Intramis	Hard to thresh because of spiny awn; taste deteriorates with longer storage.
Kapukaw	Drought susceptible.
Kurikit	Not adapted to poorer soils.
Lubang	High loss to birds because of small, smooth grain; loss to chickens because of short plant; shatters; taste deteriorates with longer storage; hard left-overs.
Luksomama	Poor taste; lodges; late maturing; susceptible to pests because of aroma at vegetative stage.
Malan	Susceptible to neck blast.
Pamintana	Not adapted to poorer soils.
Pilit Stripe	Stickiness lost with longer storage; large husk and small grain.
Pilit Tapul	Not sticky enough

Other upland rice cultivars planted in Claveria included Azucena, Kayatan, Lubang, and Mimis. Many farmers (and especially settlers from Bukidnon) had tested Dinorado, but stopped planting it because of high losses to birds. Farmers had observed that Dinorado's slightly earlier (about one week) grain development compared with Speaker, combined with small grain, led to high losses to birds. A few farmers planted a variety tested locally by IRRI (IR30716), an Indica with Japonica parent lines (M. Arraudeau, 1994, personal communication).

Farmers both in Bukidnon and Claveria wanted upland rice to have high yield, early maturation (about 120 days duration), good eating quality (however defined), resistance to lodging, and disease resistance (Table 3.4). Smaller numbers of farmers also mentioned drought tolerance or resistance, and minimal losses to birds (i.e., through erect flag leaves, large grain size, and not too early

Table 3.3. Farmers' (n = 35) upland rice cultivars and evaluations, Claveria, Misamis Oriental, 1992 wet season

Characteristics	*Rice cultivars*								
	UPLRi5	*Speaker*	*Azucena*	*Kayatan*	*Lubang*	*Mimis*	*Milagrosa*	*IR30716*	*Others*[a]
Farmers growing cult. (%)	51	54	6	9	6	9	6	6	17
Farmers dropping cult. (%)	11	26	11	14	0	9	3	11	66[b]
Mean area grown (ha)	0.84	0.65	0.50	0.58	0.50	0.26	0.75	0.82	0.31
Positive characteristics:									
good taste/eating quality	89	64	100	100	0	100	0	100	0
good expansion cooked	11	16	0	0	0	0	0	50	0
when cooked stores well	6	5	0	0	0	33	0	0	0
high tillering	94	53	50	67	100	0	0	100	0
does not lodge	83	16	0	33	50	0	0	100	0
high yield	78	68	100	67	50	100	0	100	7
high milling recovery	72	63	50	33	50	0	0	100	0
some drought resistance	61	21	0	33	0	33	0	0	0
resistance to birds[c]	56	21	0	0	0	0	0	50	0
does not shatter	50	37	100	67	100	0	0	0	0
pest and disease resistance	50	63	0	67	0	33	0	100	0
responsive to fertilizer	44	0	0	0	0	0	0	0	0
heavy grain	22	5	0	67	100	33	0	0	0
tolerates acid/poor soil	22	53	50	67	0	33	0	0	0
good panicle extension	22	5	50	67	0	33	0	0	0
early maturing	17	0	50	0	50	0	0	50	0
high market price	0	26	0	0	0	0	0	0	3
suppresses weeds	0	21	0	0	0	67	0	0	0
easy to thresh	0	16	0	0	0	33	0	50	0
long panicle	0	16	50	67	0	33	0	0	17
Negative characteristics:									
late maturing	22	26	0	100	0	0	0	0	17

Table 3.3. Farmers' (n = 35) upland rice cultivars and evaluations, Claveria, Misamis Oriental, 1992 wet season. (continued)

Characteristics	*Rice cultivars*								
	UPLRi5	*Speaker*	*Azucena*	*Kayatan*	*Lubang*	*Mimis*	*Milagrosa*	*IR30716*	*Others*[a]
pest and disease problems	22	16	50	0	0	33	0	0	50
short: chicken losses	17	0	0	0	0	0	0	0	0
more weeds (architecture)	17	0	0	0	0	0	0	0	0
shatters	17	14	0	0	0	100	0	50	17
lodging	0	46	100	100	0	100	0	0	50
bad taste	0	39	0	0	0	0	0	50	0
drought intolerance	0	21	50	0	0	33	0	0	0
susceptible to birds	0	21	0	100	0	100	0	0	83
low yield	0	0	50	33	0	0	0	0	0
low milling recovery	0	0	50	0	0	0	0	50	17
small grain	0	0	0	33	0	0	0	0	17

a. Basura, Bucay, Dinorado, IAC25, Dahili, Kabuyok, Kuhitan, and Pilit were each grown by only one farmer; many farmers had planted Dinorado but had dropped the cultivar because of bird losses.
b. Farmers had also planted and dropped Karagupan, Maria Gakit, and Kurikit.
c. Because of an erect flag leaf in the case of UPLRi5 and large grain in the case of Speaker.

Table 3.4. Characteristics of upland rice cultivars desired by farmers, Bukidnon (n = 20) and Claveria (n = 24), Philippines, 1992 wet season

Desired characteristics	*Farmers naming characteristic (%)*	
	Bukidnon	*Claveria*
High yield	90	100
Early maturation	95	79
Good eating quality	85	88
Does not lodge	65	54
Disease resistance	40	38
Drought tolerant/resistant	25	13
Does not shatter	15	4
Not attractive to birds	15	17
Suppress/tolerate weeds	15	4
High tillering	10	8
Easy to thresh	5	4
Heavy grain	5	0
High milling recovery	5	4
Long panicles	5	4
Compatible with corn grits	5	0
High price	0	13
Adapted to various soils	0	21
Even maturation	0	4

grain development). Many farmers also offered more unusual desirable characteristics: a farmer in Bukidnon wanted a rice that could be eaten with corn grits.

Rainfed lowland rice

Farmers of rainfed lowland rice in Solana planted both MVs and traditional cultivars (Table 3.5): 28 per cent grew Wagwag Fino, 28 per cent grew Ramsey, 19 per cent grew an MV; and 18 per cent grew Java. Traditional cultivars were valued for good eating quality, high market price (Wagwag), drought tolerance (Malagkit, Wagwag, and Ramsey), submergence tolerance (Java), and for reasonable yields (2.0–2.3 t/ha) at low input levels. Almost half the farmers growing Wagwag Fino reported late maturation as a problem. The MVs were planted in favourable areas and included IR36, IR60, IR66, or IR68. These were favoured for high yield and early maturation, but disliked for poor eating quality and the perceived need for high input use levels.

Overall, Solana farmers sought good eating quality, high yield, flood and drought tolerance, and low-input requiring cultivars (Table 3.6). These farmers had discontinued different traditional cultivars for a range of reasons (e.g., tall stature, poor crop stand, difficult to harvest and thresh), and had dropped MVs mainly because they were seen as needing irrigation and expensive inputs. Solana farmers matched their traditional rices to different parts of their rainfed lowland environment, and matched the MVs to limited areas with good water control.

In Tarlac, farmers grew a mix of MVs and traditional cultivars (Table 3.7): 78 per cent planted one of a number of MVs; 30 per cent grew Benser; and smaller proportions of farmers grew Okinawa, Pagay Iloko, Hamog, and others. The MVs were valued for high yield and early maturation, but disliked for the need

Table 3.5. Farmers' (n = 32) rainfed lowland rice cultivars and evaluations of cultivars, Solana, Philippines, 1992

Characteristics	*Cultivars*						
	Improved	*Wagwag Fino*	*Ramsey*	*Java*	*Raminad*	*Malagkit*	*Others*
Farmers growing cult. (%)	19	28	28	18	9	9	19
Mean area (ha)	1.6	1.1	0.9	1.2	0.5	0.3	1.3
Yield (t ha-1)	n.a.	2.3	2.4	2.0	1.9	0.5	1.5
Positive characteristics:							
good taste/eating quality	50	100	100	67	100	100	80
high market price	0	67	0	0	0	0	0
expands when cooked	50	56	33	50	33	67	40
drought tolerant	17	56	56	50	33	67	40
pest and disease resistant	0	56	22	17	0	33	20
early maturing	100	44	0	0	100	0	20
no inputs required	0	44	44	67	0	33	40
more, long, heavy grain	0	33	22	50	0	0	40
for delicacies	0	33	11	0	0	100	20
easy to thresh	50	22	22	17	67	0	60
high milling recovery	17	22	0	33	0	0	20
high yield	100	11	11	0	33	0	0
more productive tillers	83	11	33	0	0	0	0
uniform maturing	0	11	0	0	0	0	0
large panicles	17	11	0	0	0	0	0
tolerates/competes with weeds	0	11	22	50	0	0	60
flood tolerant	50	0	22	100	67	0	20
tall stature	0	0	22	33	0	0	0
good for nonwaterlogged area	0	0	22	0	0	0	20
soft left-overs	0	0	11	0	0	0	20
does not lodge	0	0	11	17	0	0	0
medium height	0	0	0	0	33	0	0
medium maturing	0	0	0	0	0	0	20
not attacked by birds	0	0	0	0	0	33	0
does not shatter	0	0	0	17	0	0	0

Table 3.5. Farmers' (n = 32) rainfed lowland rice cultivars and evaluations of cultivars, Solana, Philippines, 1992 (continued)

Characteristics	*Cultivars*						
	Improved	*Wagwag Fino*	*Ramsey*	*Java*	*Raminad*	*Malagkit*	*Others*
Negative characteristics:							
late maturing	0	44	0	11	0	0	0
insect and disease problems	33	22	0	0	0	33	0
low tillering	0	22	11	0	0	0	0
lodging	50	22	0	11	0	0	0
less drought tolerant	17	11	0	0	0	0	0
shatters	0	11	11	0	0	0	0
requires high input levels	83	0	0	0	33	0	20
low grain weight	0	0	0	6	0	0	0
difficult to thresh	0	0	0	11	0	0	0
early maturing	0	0	0	0	0	0	20
poor eating quality	50	0	0	0	0	0	0
none	17	0	44	0	33	0	0

Table 3.6. Rice cultivar characteristics desired by farmers, Solana, Philippines, 1992

Characteristic	*Farmers (%)*
Good quality eating/aromatic	41
High yield	28
Flood tolerance	22
Low input levels required	19
Drought tolerance	19
Long, heavy panicles	16
Early maturing	16

for high input levels and good water control. Traditional cultivars were desired for good eating quality (although not all traditionals), tall stature and flood tolerance, late maturation, and low input requirements (Table 3.8). As in Solana, Tarlac farmers planted the MVs in areas having good water control or supplemental irrigation, and matched their traditional cultivars to lower areas where submergence was common.

Discussion and conclusions

The 'Green Revolution', in terms of developing, introducing, and adopting suitable new varieties, bypassed the less-favourable rice environments – or less-favourable areas within particular regions. Traditional cultivars dominate the uplands of Bukidnon; and traditional and improved or modern varieties dominate in Claveria (upland) and Solana (rainfed lowland). Although Tarlac farmers planted MVs on portions of their rainfed lowland areas having good water control, they had to continue to rely upon traditional cultivars in the remaining areas.

Farmer adoption of successful traditional varieties can lead to loss of local agrobiodiversity. Examples in which one or two traditional rice cultivars came to dominate (in terms of numbers of farmers planting and area planted) are Dinorado in Bukidnon, Speaker and UPLRi5 in Claveria, and Wagwag Fino and Ramsey in Solana. The maintaining of a range of traditional rices, but with emphasis on a few has been encountered in other upland rice areas: in Laos (Fujisaka, 1991), Myanmar (Fujisaka *et al.*, 1992), and Indonesia (Fujisaka *et al.*, 1991b). Similarly, in the rainfed lowlands, a few dominant cultivars among many often correspond to farmers' matching of cultivars to lower flood-prone, and middle and upper drought-prone terraces (Fujisaka, 1990, Fujisaka *et al.*, 1991a).

Cultivars are needed that match the local diversity of unfavourable environments. All farmers do not grow Dinorado in the areas surveyed in Bukidnon, apparently because of slight differences in elevation and temperature. Many farmers tested Dinorado in Claveria (many Claveria settlers came from Bukidnon), but the cultivar was less suited to the area because of high losses to birds. Farmers in Solana and Tarlac plant MVs where possible, but continue to match traditionals with different hydrological conditions on less favourable lands.

Farmers' and rice breeders' main selection criteria are more similar than different, but are not completely congruent. Combined sets of criteria are ideally

Table 3.7. Farmers' (n = 33) rainfed lowland rice cultivars and evaluations of cultivars, Tarlac, Philippines, 1992

Characteristics	*Cultivars*				
	Improved[a]	*Benser*	*Okinawa*	*Pagay Iloko*	*Hamog*[b]
Farmers growing cultivar (%)	78	30	21	18	12
Mean area (ha)	1.3	1.5	0.8	1.2	3.0
Mean yield (t ha-1)	4.2	3.0	2.1	1.6	1.8
Positive characteristics:					
high yield	88	60	14	0	25
early maturing	50	0	0	0	0
soft when cooked	42	0	0	60	0
good taste/eating quality	31	100	28	100	100
expands when cooked	0	70	28	0	75
for delicacies	0	30	56	0	25
white grain	0	20	56	83	50
tall stature	0	100	100	100	100
flood tolerant	0	100	86	100	25
late maturity	0	100	100	100	100
low inputs required	0	50	100	100	100
pest and disease resistance	7	90	14	83	25
weed tolerant	0	30	71	83	75
drought tolerant	0	50	14	0	50
long grain	0	40	0	0	25
high market price	0	0	0	0	25
Negative characteristics:					
needs high input levels	100	0	0	16	0
needs good water control	57	0	0	0	0
hard when cooked	15	50	100	0	0
lodging	15	30	71	0	50
poor eating quality	12	0	56	50	0
susceptible to drought	7	0	0	33	0
difficult to thresh	0	50	0	50	0
low yield	0	0	71	50	75
low milling recovery	0	0	0	83	0

a. IR4, IR20, IR36, IR42, IR48, IR59, IR60, IR64, IR66, IR70, BPIR10, and Super.
b. A few farmers also planted Pinarya, Lamio, Dinaloson, Magsopa, Numero Doce, Azuceña, and Los Baños.

Table 3.8. Farmers' (n = 33) desired traditional rice cultivar characteristics, Tarlac, Philippines, 1992

Characteristic	*Farmers choice (%)*
Good eating quality/aromatic	100
Late maturing	54
Tall stature	48
High yield	36
Expands when cooked	18
White	12
Flood tolerance	9
Low input levels required	9

suited to newer breeding strategies where international agricultural research centres provide donor lines featuring major resistance or adaptability, and national programmes make more specific crosses and selections to develop locally appropriate varieties. Both upland rice farmers and breeders seek high yields, adaptability to local soils, medium duration (not so short that the crop is unstable in the face of stresses), lodging resistance, high milling recovery, and pest and disease resistance. Farmers, however, also consider traits such as taste, maintenance of taste in stored grain, hardness of cooked left-overs, attractiveness to birds and chickens, and the difficulty of handling awned rices. Farmers in the rainfed lowlands, who face long periods of flooding, chose cultivars that are tall and purposely of long duration (i.e., photoperiod sensitive) – characteristics quite unlike the short-statured and short-duration, Green Revolution rices.

Farmers' knowledge and selection criteria related to rice cultivars are substantially 'scientifically' sound. For example, farmers are correct in that plant architecture, grain size, and relative timing of grain development influence bird losses; that plant architecture influences weed competitiveness; that aromatic plants (e.g., Speaker) can be relatively more attractive to some insect pests; that some cultivars are more tolerant of acidic or poor soils; and that sharp awns are difficult to handle.

Providing a 'Green Revolution' in the uplands and rainfed lowlands may be difficult precisely because farmers make sound choices, practice sound management, and are successful to the degree that risks inherent in the system allow. In the 1988 wet season, a good year in Claveria in terms of rainfall, farmers' traditional varieties on their own fields yielded a mean 1.9 t/ha; UPLRi5 without fertilizers yielded a mean 2.8 t/ha; and UPLRi5 with small amounts of inorganic fertilizer (25kg N, 6kg P2O5, and 5kg P2O) yielded 3.2 t/ha (Fujisaka, 1991). Conversely, in 1991, a drought year in Claveria, all rice fields of upland farmers yielded a mean 0.8 t/ha (Fujisaka, 1993).

The uplands and rainfed lowlands pose substantial challenges and offer significant opportunities for problem-solving research. 'Modern' varieties, which address the diverse, risky conditions inherent in these environments, and which are clearly superior to traditional cultivars, have not yet been developed. The opportunities lie in the diversity of traditional rices that farmers have selected over time for suitability to local conditions, and in understanding and applying the knowledge underlying these selections.

4. Environmental dynamics, adaptation, and experimentation in indigenous Sudanese water harvesting

DAVID NIEMEIJER

Abstract

SOME 2000 YEARS ago the Beja of North-eastern Sudan first engaged in runoff farming as a side-line to their mainly pastoral life. At present it has become a vital source of subsistence for several of the Beja ethnic groups. The best-watered soils were lost to large-scale irrigation schemes and the traditional desert trade was supplanted by modern means of transportation. On the arid piedmont plains to the east and north of Kassala town the Hadendoa and Beni Amer peoples, two of the Beja ethnic groups, engage in three forms of indigenous runoff farming: wild flooding, water spreading, and *teras* water harvesting. The highly dynamic environment with low and irregular summer rainfall and continually shifting wadi courses is hard to master. Nevertheless, the local agro-pastoralists have succeeded in developing highly adapted forms of these indigenous runoff farming systems. The key to their long-term survival is the degree to which they have been successful in coping with environmental change and dynamics. As wadi courses shift and rainfall varies, the Hadendoa and Beni Amer are able to attain a reasonable and sustainable harvest only through continued experimentation. Especially among the Hadendoa this has led to extremely varied shapes and dimensions in their water-harvesting structures, as each farmer attempts to cope successfully with the constraints and opportunities that a specific site offers in the course of time. For the Beja, as for many other agriculturists, experimentation is the key to adaptation – a fact that, once recognized, has important implications for attempts at technology transfer.

Introduction

Long before the onset of agriculture the survival of humankind depended on adaptation and experimentation: adaptation to the dangers and opportunities of the environment and experimentation with tools and techniques in order to further master that environment.[1] Not only did adaptation and experimentation eventually lead to the development of agriculture, but these characteristics continue to be among the vital forces that support agricultural livelihoods in our ever-changing world.

Leakey (1936) was among the first to recognize the well adapted nature of African agriculture as he encountered it among the Kikuyu. In his view, local methods should not be replaced by (less adapted) European ones. Still, he suggested, they could be improved through scientific research, for he believed scientific research to be more powerful than what he called the trial and error research of the Kikuyu. In mentioning that 'the methods of research by trial and error have led the Kikuyu to important conclusions' he nevertheless acknowledged the value of farmer experimentation (Leakey, 1936).

In the early 1970s Johnson (1972) drew attention to indigenous experimentation again, and maintained that individual differences in agricultural practice

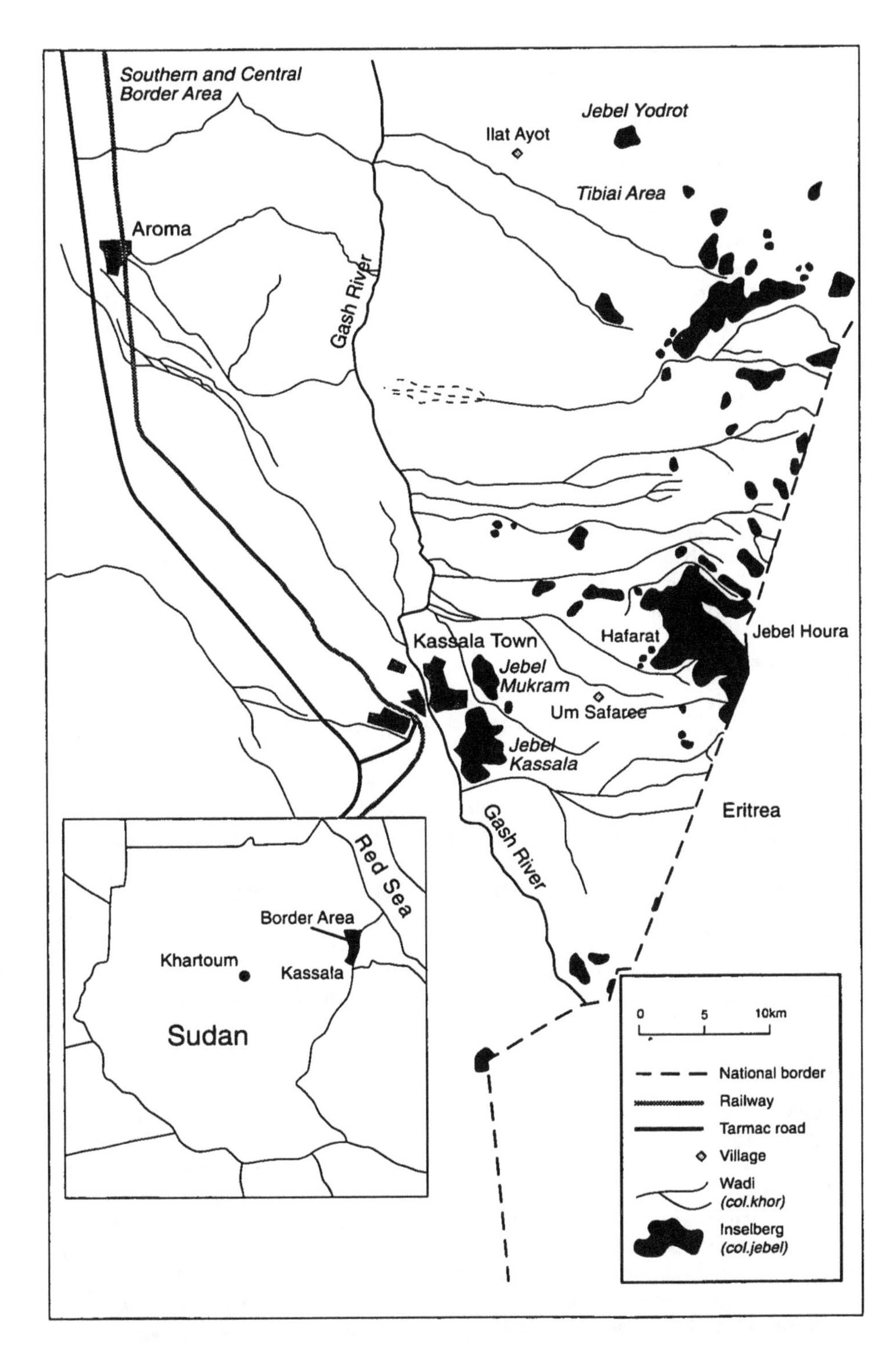

Figure 4.1 Main features of the southern and central Border Area; adapted from Van Dam and Houtkamp (1992) and Westhoff (1985)

and systematic experimentation were probably pervasive in traditional agricultural communities and an essential component of their adaptive processes; a view that has since gained increasing power especially among anthropologists (see for example Richards, 1985), and has led to a series of new publications starting in the early 1980s. Much of the recent material on farmer experimentation (for example Rhoades and Bebbington, 1988; Brouwers, 1994; Dangbégnon 1994) focuses on agronomic measures, new crops, and cultivar selection and breeding (especially with the current interest in biodiversity). In contrast, this chapter will focus on experimentation in another field, namely that of ethno-engineering. Ethno-engineering is a term that is sometimes used to denote physical structures that local farmers build to serve field crop production (Reij, 1991). Irrigation canals, or soil-conserving stone lines are good examples of such ethno-engineering practices, as are the brushwood and earth bunds of the Sudanese Beja peoples that we will discuss here.

The material presented here is based on research among the agro-pastoral Hadendoa and Beni Amer, two of the Beja ethnic groups, that live in the Border Area of Sudan's Kassala Province (an 8600 square kilometre pie-slice bordering Eritrea and located to the north and east of Kassala town, see Figure 4.1). These agro-pastoralists employ a range of indigenous runoff farming techniques for sorghum (*Sorghum bicolor, S. vulgare*) and millet (*Pennisetum typhoideum*) cultivation. Through continued experimentation and adaptation they have developed an arable farming system that is remarkably viable and sustainable considering the harsh and dynamic environment.

Fieldwork methods

The present report is based on field research conducted in Eastern Sudan in the period October 1990 to January 1991.[2] At the time there was a major drought (with 76 mm rainfall, only 25 per cent of the long-term average, it was one of the driest years of the century) and this was one reason to focus on the engineering structures rather than on agronomic practices. It also limited the number of farmer interviews, because many farmers left the village in search of alternative sources of livelihood.

The material presented here is based on detailed field observations in several farming areas of the southern and central part of the Border Area,[3] informal interviews with key informants and farmers at field sites, level surveys of three runoff farming fields (*terus*), air photo interpretation of farming areas, soil survey work, and soil sample analysis. In total, over 30 scattered runoff farming fields were visited.

Historical background[4]

Somewhere between 4000 BC and 2500 BC the Beja crossed the Red Sea from Arabia and settled in Sudan's Eastern Desert (Paul, 1954). From what is currently known, livestock-keeping has been the culturally preferred mode of subsistence for the Beja ever since. Nevertheless, they have always engaged in various other activities. Already before the Christian era they developed exchange contacts, and by the end of the first millennium the Beja had intense trade relationships and conducted caravans through the desert, taking care of the local supplies and controlling the hinterland (Hjort af Ornäs and Dahl, 1991; Palmisano, 1991). Some 2000 years ago runoff farming gradually replaced wild grain collection[5] and probably took place mainly in the form of wadi cultivation

(Hjort af Ornäs and Dahl, 1991; Ausenda, 1987). A notable exception was the large inland delta area of the seasonal Gash river that allowed flood-recession agriculture. Ibn Hawqal, an Arabic traveller from the tenth century already noted sorghum cultivation in the Gash delta (Van Dijk, 1995; Hjort af Ornäs and Dahl, 1991), and as early as the fifteenth century Gash sorghum was a much famed export product of north-eastern Sudan (Hjort af Ornäs and Dahl, 1991).

From the literature (Newbold 1935; Paul, 1954; Hassan Mohammed Salih, 1980; Hjort af Ornäs and Dahl, 1991; Sørbø, 1991) it becomes clear that the Hadendoa and Beni Amer peoples, the two Beja ethnic groups that occupy the Border Area, the territory on which this paper focuses, traditionally lived dispersed in small family groups. These units lived a more or less sedentary life at a number of seasonal sites. Year after year they would return to the same dry and wet season sites. This scattered way of life was a successful means of coping with the scarce environmental resources that a harsh climate allowed. It guaranteed a well-balanced population distribution over the whole area. Large-scale seasonal migrations were not necessary, thus sparing the environment. Only under certain conditions (like very low rainfall) larger scale migrations took place to the Gash and Tokar deltas or to the winter rainfall area near the Red Sea. Places like the Gash delta, the Red Sea hills and the Eritrean highlands offered dry season's grazing, whereas the wet season allowed a wide dispersal of man and beast. The Beja had achieved considerable flexibility in their exploitation of a natural environment characterized by great fluctuations both in the amount and the nature of available resources, through their socio-cultural system that supported dispersed residence, and through the availability of drought-time refuges.

As happened everywhere in Africa, the way of life of the Beja was seriously affected by the establishment of colonial power in the early nineteenth century. Taxation, though certainly at first very ineffective (Hjort af Ornäs and Dahl, 1991), influenced agricultural and settlement patterns. Tax gathering expeditions had to be avoided, leading to additional movements of stock and conflicts in the tribal-hierarchy about the contributions to taxation. Conflicts over land and land use could no longer be dealt with by traditional law (and arms), which impelled overexploitation by multiple users. According to Baker (1867) taxation of cultivated land discouraged agricultural (surplus) production.

During the early years of Egyptian and Anglo-Egyptian occupation environmental effects must still have been limited. The situation changed, however, when the colonial hold on the area increased around the beginning of the twentieth century and 'development' activities were initiated. Public works and irrigation initiatives along the Gash required labourers who were at first attracted among the local Beja. In years of high rainfall or floods the Beja, however, planted their own crops – leading to a labour shortage which was counteracted by the government by the attraction of immigrant labour in the form of Eritreans, Somalis, Abyssinians and the so-called Westerners (West African Muslim pilgrims). With the creation of the Gash Delta Scheme in the early 1920s the labour requirements grew, and because of the unwillingness of the local Beja to change their lifestyle and settle as cotton farmers in the scheme, still more immigrants were attracted (McLoughlin, 1966). The Hadendoa did not like these developments and the intrusions of other ethnicities on their grounds. Even though the government decided to give more thought to the rights of the Hadendoa, the scheme remained at odds with the interests of the Hadendoa. They had to accept 'foreigners' on their land, they lost control over grazing and cultivation rights and they were supposed to get rid of their stock and settle for cotton farming. What happened was that they let others cultivate their land and eventually lost control over much of the best land. They could no longer utilize the land in their own way and to their own accord (Paul, 1954; McLoughlin, 1966; Hassan Mohammed Salih, 1980). In effect, they were to a large extent driven into the comparatively marginal areas like the Border Area. This increased the population pressure in these areas, also during the dry season when they could no longer graze their stock throughout the Gash delta, and has led to deforestation and land degradation.

A further set-back for the local Beja was the introduction of modern means of transportation and when in 1924 the railway that connected Kassala to the Red Sea coast was completed, the desert trade, which had been a virtual monopoly of the Hadendoa, dwindled (Newbold, 1935). All in all, the Hadendoa and Beni Amer became increasingly constrained in their traditional livelihood activities, as the best watered soils were occupied by large-scale irrigation schemes, and the transhumance routes were curtailed by mechanized farming in the southern pastures. It is within this context that, in the early decades of the present century, runoff farming became an increasingly important contribution to the livelihood of the population of the marginal Border Area (Niemeijer, 1993).

The runoff farming systems and their position within the landscape

Geomorphologically the Border Area may be described as piedmont plains consisting of pediments largely covered by coalescing alluvial fans and visually dominated by occasional inselbergs (rock outcrops). The average altitude of the plains is some 500 m above mean sea level, with a gentle slope (averaging less than 1 per cent) from the east to the west. The plains are dissected by numerous east–west flowing smaller and larger wadis (ephemeral streams) that drain the foothills of the Eritrean highlands, as well as the local inselbergs (see Figure 4.1). These streams spread and contract several times along their course, at points where the gradient respectively diminishes and steepens, resulting in a relatively regular east–west oriented drainage pattern (see Figure 4.2). This complex drainage pattern and the variability of discharge and sediment load has resulted in soils with an extremely high spatial variability, often leading to completely different soil profiles (in terms of texture and horizon depth) at distances of as little as 10 m. The following soils may be recognized: Vertic, Haplic, and Calcic Luvisols; Vertic Cambisols; Eutric Fluvisols; and Eutric Regosols.[6] Long-term average (summer) rainfall decreases from some 300 mm in the south to 150 mm in the northern part of the Border Area, but in recent decades the average has decreased by some 20 per cent (Niemeijer, 1993).

In many ways the Border Area offers an ideal environment for runoff farming. Rainfall is too low for regular rainfed cultivation, while the gentle slope and the relatively low population density allow for large, sparsely vegetated catchments. Initially (that is some 2000 years ago when the Beja first engaged in the cultivation of grains), wildflooding, a type of floodwater harvesting without artifacts on the alluvial flats, was probably the main form of runoff farming. It required relatively little effort in the form of land preparation, while returns could be considerable in mediocre to wet years. During colonial times, the Hadendoa and Beni Amer populations became increasingly dependent on the Border Area for their grain production, as a result of the earlier described colonial policies, and the need arose to expand the cultivated area. The areas that received water from

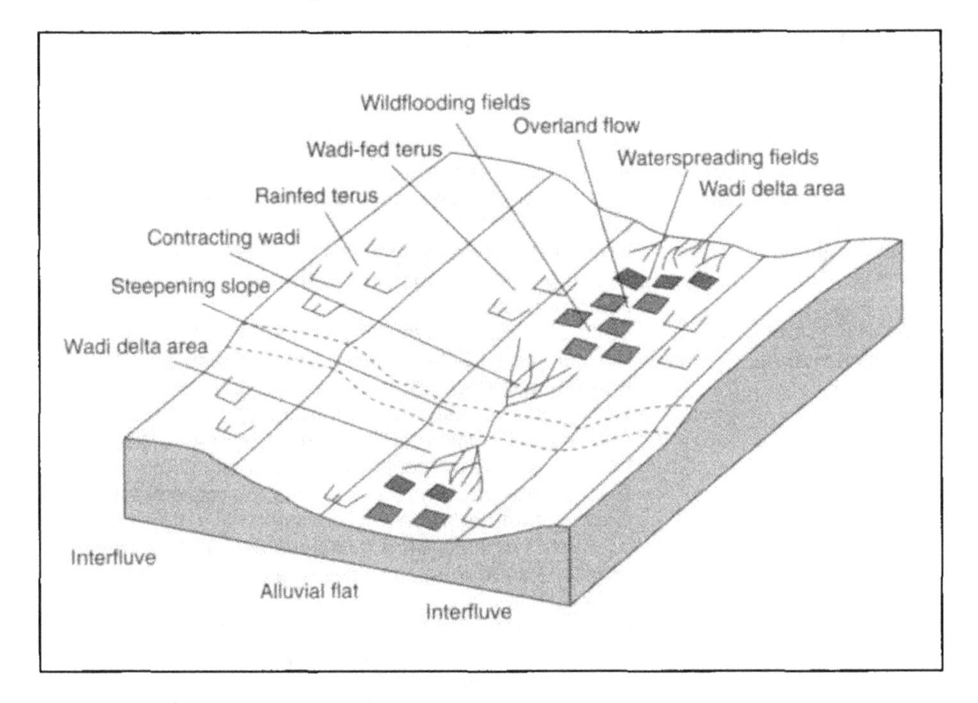

Figure 4.2 Spatial position of the Border Area runoff farming systems illustrated schematically

naturally spreading wadis were limited and it is likely that, in order to expand the cultivable area, the people at some point turned to water spreading, a type of floodwater harvesting that uses (mainly) brushwood spreaders on the alluvial flats. But, as suitable space on the alluvial flats ran out, the population soon had to turn to yet another technique.

Probably it was in the 1940s that the local population developed their own form of the *teras* system, a rainwater harvesting system that – unlike wildflooding and water spreading (both floodwater harvesting systems) – did not depend on the occurrence of stream-flow.[7] The *teras* system probably has its roots in the area of Sennar where during the Funj Kingdom (1504–1820) elementary forms of this technique were developed (Craig, 1991). Later the technique spread to large parts of (semi-)arid Sudan. Jefferson (1949) estimates that during the 1940s some 5 per cent to 15 per cent of Sudan's sorghum crop was produced on *teras* fields.[8] Presently, the *teras* system is the most widely used runoff farming system in the Border Area and covers some 40 per cent of the cultivated area (Van Dam and Houtkamp, 1992).

A *teras* consists of a field surrounded on three sides by low earth bunds while the upstream side functions as an inlet for surface runoff (see Table 4.1 for some basic statistics of the *teras* system). Figure 4.3 shows a 3D-sketch (not to scale) with the basic *teras* features. All *terus* (plural of *teras*) have a main bund consisting of a base contour bund (1) as well as two outer collection arms (2). The numerous small parallel ridges inside the *teras* are weeding ridges (3) that result from the weeding pattern (and also occur on wildflooding and water spreading fields). The inner collection arm (4) and the inner diversion arm (5) are optional features that do not occur on all *terus*. The threshing circle (6) is

Table 4.1. Basic statistics of the *teras* system of the Border Area

Dimensions	Bund height usually around 0.5 m, but old, manually constructed bunds may be over 2 m high; base bunds are generally 50–300 m in length; arms usually 20 to 100 m in length; sizes vary from 0.2 to 3 ha
Labour investment for construction	7 to 19 person-days per hectare (depending on soil type and season)
Agricultural cycle	Land preparation (May/June); sowing (June/July); weeding (June-August, up to three rounds); harvesting (October/November)
Seasonal labour requirements	66 to 80 person-days per hectare (land preparation/ maintenance 12–14; sowing 2–4; weeding 14–24; harvesting, threshing, bundling and transport 38)
Tools used	Sowing with a planting stick (generally on untilled/ unploughed soil); weeding with a hoe; harvesting with a hoe, ax, or knife; threshing with a heavy stick
Crops	Mainly sorghum (*Sorghum bicolor, S. vulgare*) and millet (*Pennisetum typhoideum*); occasionally water melons (*Citrullus Vulgaris*), okra (*Hibiscus esculentus*), rosella (*Hibiscus sabdariffa*), and sesame (*Sesamum indicum*) on the wetter parts
Plant densities	Plant spacing is variable, as is the number of plants per seeding hole (5 to 10); average plant densities around 58000 to 62000 plants/ha
Yields	300 to 600 kg/ha in mediocre to good rainfall years[9]

Source: Niemeijer (1993, 1998); Van Dam and Houtkamp (1992); Van Dijk and Mohamed Hassan Ahmed (1993).

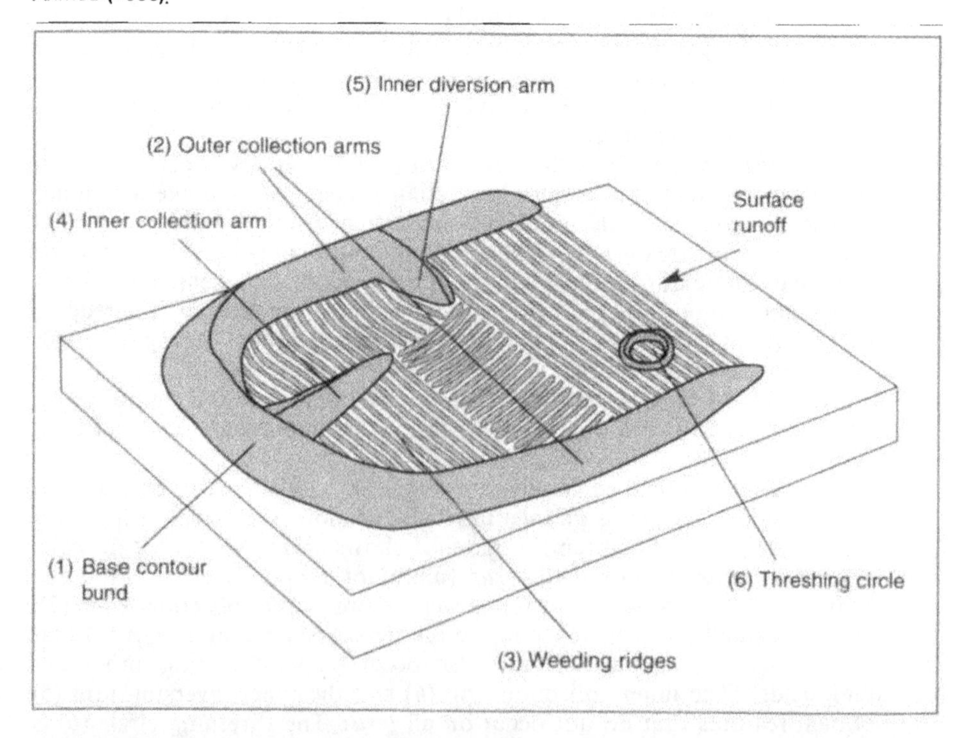

Figure 4.3 3D-sketch of a *teras* (not to scale)

only found on a few *terus* and is used to reduce grain losses during threshing. A part that has not been drawn is the shallow channel that is sometimes found on the inside of the main bund (1 & 2). This channel is the source of earth needed for heightening or repairing the main bund. In some cases it functions as an additional water reservoir for planting fruits or vegetables, usually water melons (*Citrullus Vulgaris*) or okra (*Hibiscus esculentus*). Though the *teras* technique shows local variations in size, shape and height of the bunds, the same name is used throughout the Sudan. The colloquial term *teras* is also widely used in the literature (see Tothill, 1948; Lebon, 1965; Reij *et al.*, 1988; Reij, 1991; Craig, 1991; Critchley *et al.*, 1992; Van Dijk and Mohamed Hassan Ahmed, 1993; Van Dijk, 1995) – for which reason it will also be used here.

Runoff farming in the Border Area is remarkable in the sense that in an area of only 8600 square kilometres three major runoff farming systems are in use. What is more, within the broad categories of wildflooding, water spreading and *teras* water harvesting, there is also plenty of differentiation. The wildflooding fields may be divided into those that are located within the actual streambed (wadi cultivation), and those located in areas that are naturally flooded by water that has either overflowed the river banks or flows as dispersed overland flow. Of the brushwood water-spreading system two variations may be recognized. The first uses communally built spreading structures in the main water course at the so-called project-level: fields of several farmers are receiving water from a single large structure. The second variation uses field-level spreading structures located on the individual fields. Finally, two types of *teras* systems may be distinguished. The more common system receives only overland flow (rainfed), and the other, specific to the Border Area, makes use of concentrated stream flow from one or more small ephemeral streams (wadi-fed) and is thus not a rainwater harvesting system (*pur-sang*), but a combined flood- and rainwater-harvesting system (Niemeijer, 1993).

This variety of systems is indicative of the creativity with which the local agro-pastoralists approach runoff farming. Land is not yet in short supply (calculations of Van Dijk and Mohamed Hassan Ahmed (1993) show that only some 5 per cent of the area is covered by agricultural fields) and most fields are thus located in the hydrologically favourable areas. Farmers usually have fields in several of these areas and adapt the type of runoff farming system they use to the local hydrological conditions. As Figure 4.2 shows, wildflooding is found on the alluvial flats, mainly in areas that show natural spreading of concentrated stream flow to dispersed overland flow. Water spreading is also found on these alluvial flats, but in those areas where the water flow is still too concentrated to allow wildflooding. The *teras* system occurs in intermediate positions and on the interfluves (only a few metres higher than the alluvial flats). Those located in the intermediate positions make use of both overland and stream flow, whereas those on the interfluves depend exclusively on overland flow.

The spreading structures and bunds used in the water spreading and *teras* systems are by no means uniform in size, material, shape and orientation. They could not be, for in the harsh and dynamic Border Area environment farmers need to adapt their systems optimally to the site-specific hydrology and soils. In fact, adaptation alone is not enough, for wadis and overland flow patterns shift from year to year, as do the amount of rainfall and discharges. With a coefficient of variation of 30 per cent there are large inter-annual fluctuations in the amount and timing of rainfall and, as is common in such arid environments, rainfall spatial variability is high as well. Farmers need to experiment continually to adapt their structures and planting patterns to these environmental dynamics and secure an adequate but non-destructive supply of water to their crops. How this occurs is best seen in the *teras* system.

Adaptation and experimentation in the *teras* system

In the most southern part of the Border Area, that is inhabited by the Beni Amer, most of the variation in the *terus* occurs in terms of the size of the fields and the height and orientation of the bunds. Here most of the *terus* are rectangular in shape, unlike in the central part of the Border Area occupied by the Hadendoa, where, in addition, shape is a major variable. That was one of the reasons why the farmlands of the village of Ilat Ayot, which is located in this central area (see Figure 4.1), were studied in most detail and are the source of most of the examples below (though a lot also pertains to *terus* in the other parts of the Border Area).

Regulating the water supply to the teras

Regulation of the water supply is essential to successful cropping. The farmers have to balance the water supply in such a way as to obtain the highest chance of a good harvest in most years, while eliminating the risk of destruction of their *terus* in years with high rainfall and large wadi discharges. In the Border Area environment with low irregular rainfall and dynamically changing wadi courses choice of location is vital, since it will determine the chances of success in the years to come (labour investments for the initial construction of a *teras* are high; see Table 4.1).

As can be seen in Figure 4.4 (a detail of a tracing of a 1986 aerial photograph of an Ilat Ayot farming area), *terus* are not located in the immediate stream course of a wadi. A frontal assault of the water during one of the few discharge events would certainly wreak havoc on a *teras*, and damage the bunds severely. So, instead, the *terus* are carefully constructed in those places that do receive additional wadi water, but that do not receive the full force of the discharge. Broadly speaking, two kinds of situations can be observed. Some *terus* (a and e in Figure 4.4) are positioned in those areas where the water of the wadis spreads out, leaves the defined water course and continues to flow dispersed as low kinetic energy overland flow. Other *terus* (d in Figure 4.4) are constructed on the banks of the wadis and, on discharge events, receive the low kinetic energy overflow from the only slightly incised wadis. Certain *terus* (b and c in Figure 4.4), make use of a combination of these kinds of additional ephemeral stream water.

Once the location has been fixed a lot depends on the adaptation of size, orientation, and shape to the hydrology and soils of that particular site. Through continuous experimentation the farmer tries to cope with the environmental dynamics. Shifts in a wadi's course or fluctuations in discharge frequently warrant modifications of the *teras* shape. During the yearly maintenance, and sometimes even within a single season, the farmer may modify the height, shape and extent of the various bunds, based on his[10] experiences from the previous years. The amount of harvested water may be regulated by extending the outer collection arms or changing their shape. Some *terus* are very broad, with a long base contour bund, but have very short outer collection arms; others are very deep and have a short base contour bund and long outer collection arms. The bunds are usually irregularly shaped and sometimes the size of the outer collection

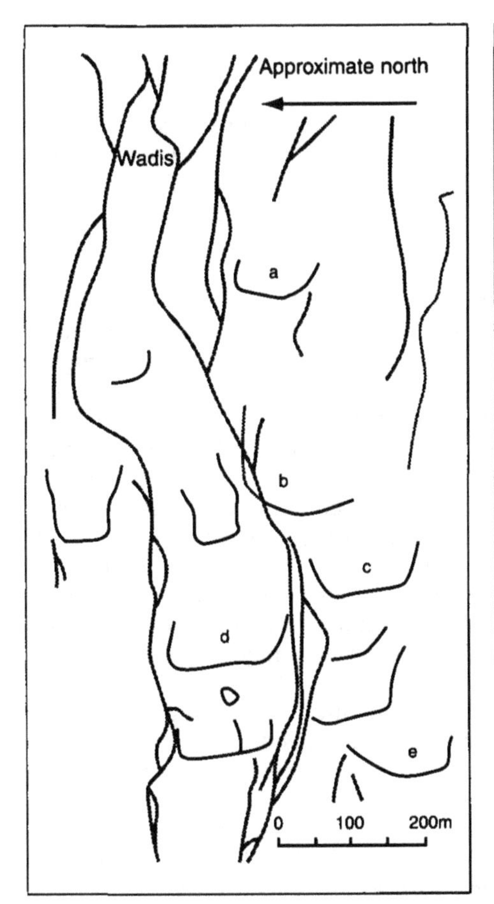

Figure 4.4 Environmental position of some *terus* (tracing of 1986 aerial photograph)

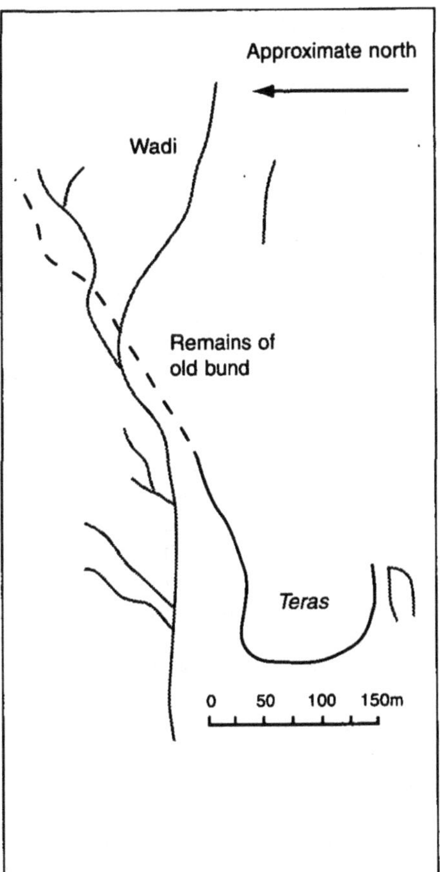

Figure 4.5 Extremely long outer collection arm of a *teras* (tracing of 1986 aerial photograph

arms of a single *teras* may differ considerably. The size of *terus* in this particular farming area ranges from 0.30 to 1.85 hectares.

One of the Ilat Ayot farmers really went to extremes to control the water supply of his *teras*. Figure 4.5 shows what now remains of an extension of one of the outer collection arms of this particular *teras*. At one time the outer collection arm had a total length of some 430 metres, of which at present some 140 metres still remain in good condition. With this long outer collection arm the farmer tried to harvest additional stream flow from an ephemeral stream. He misjudged the power of this stream and was unable to maintain this extensive outer collection arm on the long run.

Regulating the water distribution within the teras

The shape and location of the *teras* determines the approximate amount of water that is collected in that *teras*, as well as the proportions contributed by local overland flow and concentrated stream flow (wadis). Further management techniques determine the water distribution within the *teras* and are vital to guarantee a good harvest in all parts of the field, but also to prevent damage from an excess of water in a single part.[11]

The most common type of internal structure is the inner collection arm (Figure 4.3 (4)). It is used to take care of a good distribution of water within the *teras*. In some cases it prevents a very small amount of runoff being spread over the whole *teras*, whereby none of the parts receives enough water to grow anything. In other cases it guarantees that some water remains behind in slightly elevated parts of the *teras*, because the inner collection arm prevents it from all flowing to the lowest corner of the *teras*. Another, less common, type of water regulating bund is the inner diversion arm (Figure 4.3 (5)). Its main function is probably to protect against damage the part of the *teras* that receives the highest discharge from the incoming wadi, as part of the inflow is diverted to the other side of the *teras* and the force of the water is reduced.

The most extreme regulating structure is the 'mother' and 'child' structure. Inside one large *teras* (mother), one or more smaller *terus* (children) are constructed (see Figure 4.7). This technique is not so common in the Border Area (in contrast to the larger *terus* on the Clay Plains to the west of Kassala). Its function is similar to that of the inner collection arm. In wet years it allows the planted area to be larger, since runoff water is divided into separate sectors. In dry years the 'child' *teras*, because of its the smaller size, requires less runoff water to grow a crop, than the large 'mother' *teras* that may not receive enough water in proportion to its area.

If the inner arms and the mother–child structures can be regarded as the macro-structures that regulate the water distribution inside the *teras*, the direction of the weeding ridges can be regarded as the micro-structures used to distribute the life-giving liquid.

All *terus* in the Ilat Ayot area are still hand tilled. Sowing is done with a planting stick and if it were not for the weeding activities, the *terus* fields would have a smooth surface. Weeding is done with a hoe, and together with weeds some earth is scraped aside, resulting in small continuous ridges between the planting rows. The orientation of these ridges is changed by changing the direction followed during weeding. From year to year weeding directions are usually maintained (except when changed discharges require alteration) and even after several years of fallow the weeding ridges can still be recognized and form a consistent water regulating structure. The orientation of the weeding ridges is adjusted to distribute water optimally over the field. Usually the weeding ridges are kept parallel to the flow direction to guide water into the *teras*. In some cases, however, the current of incoming water is too strong and weeding ridges are created at a right angle to the inflow direction to reduce the kinetic energy of the water by increasing friction. At the same time these ridges prevent water from leaving the *teras* again.

Another situation in which the farmers change the weeding direction is where water is spread unequally over the land due to height differences or wadi characteristics. This is the case in Figure 4.6, which shows a *teras* and a part of its catchment. Only one part of this *teras* receives wadi water, resulting in a local excess of water. The farmer has therefore oriented the weeding ridges in such a way that excess water is guided towards the other part of the *teras*. This other part of the *teras* has weeding ridges in the normal direction, parallel to the inflow of local surface runoff, to enhance harvesting.

Figure 4.7 shows another example of a farmer applying multiple weeding

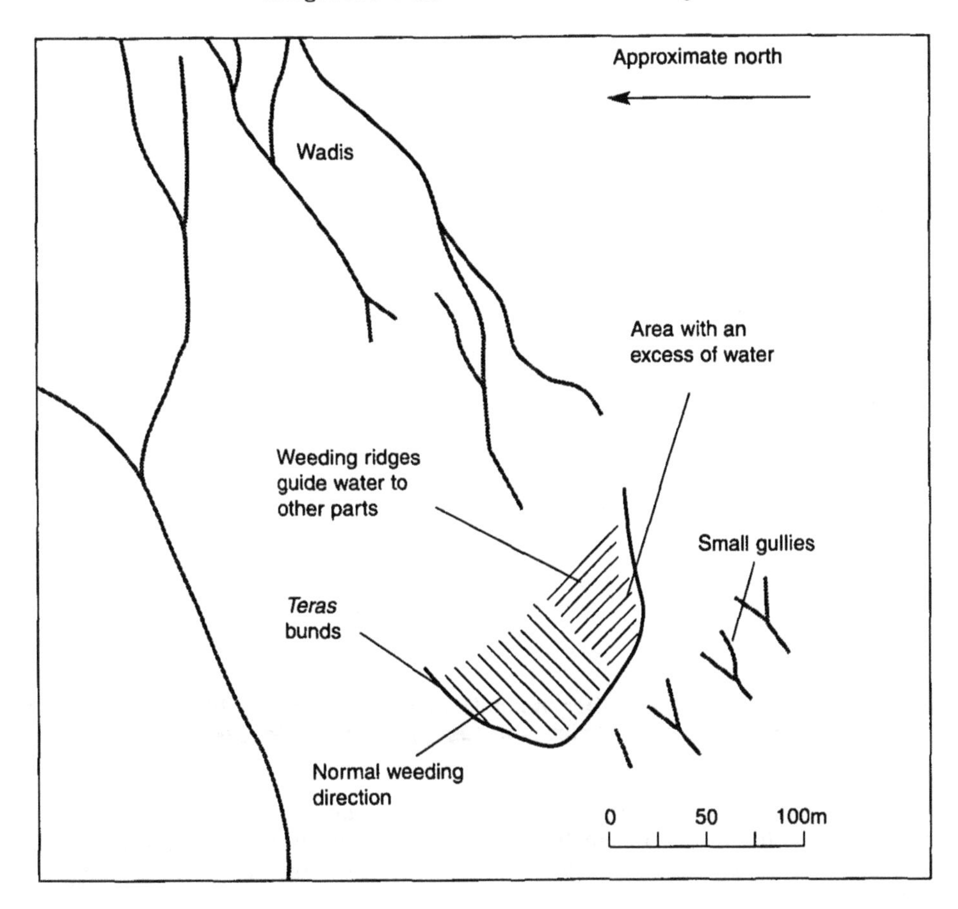

Figure 4.6 Regulating the water supply by adaptation of the weeding ridges (tracing of 1986 aerial photograph; weeding ridges not to scale)

directions to regulate the water distribution within his *teras*. In addition, this *teras* illustrates how the inner collection arm may be used. In the mother *teras* water of two wadis is used separately, one wadi supplies one side of the *teras* and the other supplies the other side. In this way the farmer enhances the chance of a successful crop in at least a part of the field if one of the wadis fails.

Individuality, adaptation and experimentation

From the above it is clear that the *terus* are highly adapted to the Border Area environment in general, but also to the specific dynamics of particular sites. The farmers cope with these dynamics by employing a number of techniques that regulate the water supply as well as the distribution within the *teras*. There is a high degree of individuality as each farmer makes his choice from the common set of techniques and adapts them to his requirements and the conditions of his particular location. This is an all but easy task in an area that recently received less than 250 mm of rainfall in eight out of ten years and where wadi courses shift from year to year. Under such circumstances experimentation is not a risk (as it is often regarded in the literature), but a necessity. Through

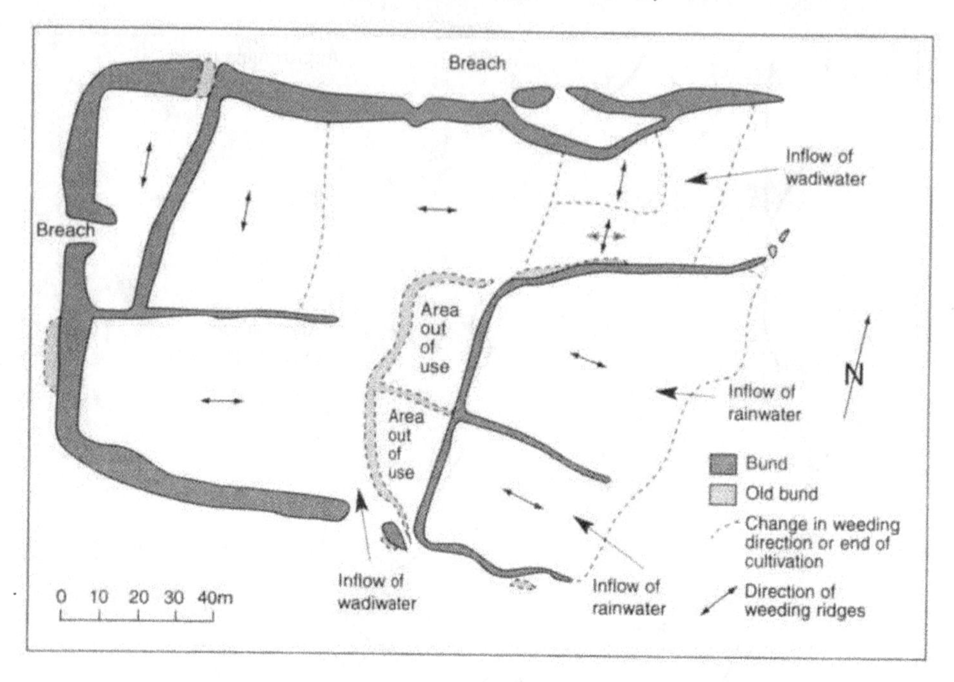

Figure 4.7 A *teras* with a mother and child structure. Weeding ridges are identical by double sided arrows (grid survey in 1990; 20 x 20m gridcells)

experimentation the farmer gains the necessary understanding of the environment, as well as of the hydraulics of his engineering structures, that is prerequisite to a successful (re-)adaptation of his *teras* to the ever-changing circumstances.

Some of these experiments are successful and are picked up by other farmers. Others, like the long outer collection arm shown in Figure 4.5, are less successful and remain a very individual, one-off experiment. There are more examples of farmers who go beyond the regular experiments and try something really new. One Ilat Ayot farmer built a *teras* within the stream course of a major wadi. According to a key informant the seeds on this *teras* are not washed away by the flood, but the *teras* bunds must be rebuilt every year: 'This requires more work than the other *terus*, but yields are higher, because there is more water from the wadi, and also from deep in the soil.' Another farmer designed curved weeding ridges that guide water from a central water source to the outer sides of the *teras*. Some farmers use branches and in a few cases (where they are available) stones to strengthen the *teras* earth bunds. Apart from these engineering experiments farmers also experiment with different sorghum and millet varieties and non-standard crops, like okra or water melon, in part of their *teras*. There are about eight different sorghum varieties in the area, each with different characteristics: e.g., drought-tolerance, taste, growing season, market value. Some farmers plant up to three different varieties on a single field to optimally benefit from soil moisture differences and to minimize risk due to variable rainfall. Plant spacing is varied from year to year (Van Dijk and Mohamed Hassan Ahmed, 1993) and there are even indications that some farmers adapt plant densities relative to the expected soil moisture gradient within their field (Niemeijer, 1993). Finally, there are also long-term changes that require re-adaptation of runoff farming techniques. The disappearance of trees in much of the southern part of the Border Area has, for instance, led to experiments with new kinds of water spreading structures that do not use brushwood, but are based on low broad-earth bunds instead.

Viability and sustainability

An important question is whether this adaptation and experimentation has led to a viable and sustainable farming system. If we look at grain yields alone the results do not seem very impressive. Grain yields under the *teras* system range from 0 in drought years to some 300 kg/ha in mediocre rainfall years, and up to 600 kg/ha in good rainfall years (Van Dam and Houtkamp, 1992; Van Dijk and Mohamed Hassan Ahmed, 1993). Under the socio-economic conditions prevalent in the area, the stalk yield, however, represents an economic value comparable to that of the grain yield (because it is used for two to six months of animal fodder (Van Dijk, 1991) and as construction material). Calculations based on figures for the southern Border Area in the mediocre year 1983 (Van Dijk, 1991; Van Dam and Houtkamp, 1992), for example, show that the monetary value of the average grain yield (340 kg/ha) and the average stalk yield (270 bundles/ha) were equal (each approximately 190 Sudanese Pounds). Some simple calculations in Niemeijer (1993) indicate that in mediocre to good rainfall years arable farming (*teras*, wildflooding, and water spreading together) is a viable labour investment – generating enough food to feed the average family, as long as some stalks are sold to buy additional grain. Good years, in addition, allow for monetary profits. These findings are in agreement with Van Dam and Houtkamp (1992), whose respondents indicated that arable farming contributed on average 70 to 80 per cent to household subsistence.[12] Even in those years when the grain yield fails, the stalk yield still covers part of the labour investments, a fact that is of vital importance in this unpredictable environment. All in all, runoff farming forms an important and viable component in the diversified livelihood strategies of the local population.

Also, in light of the aridity of the environment and the sparse natural vegetation the crop yields are, in fact, not that low. Certainly, if we consider that cultivation is almost continuous (generally, fallow occurs only in the absence of rain), the yields are actually quite an accomplishment; an accomplishment largely due to the fact that these water harvesting systems also harvest nutrients. While the wild flooding and water spreading systems owe their yearly nutrient input mainly to their position on the flood plain, the nutrients harvested in the *teras* system are largely a result of the human-made structures that retain organic matter, silt, and fertile waters. Again, the position and shape of the bunds determine the amount of material that is harvested, while the weeding ridges contribute to the distribution of nutrients over the whole *teras*. This way, up to a few centimetres of nutrient-rich sediment is harvested on a yearly basis. The fact that weeds and stubbles are not removed from the fields further provides for sustained soil fertility (Niemeijer, 1993; 1998).

Water harvesting efficiency of the runoff farming systems was (due to the drought at the time of research) not measured, but considering the fact that with current rainfall trends ordinary rainfed farming would be impossible in eight out of ten years (years with rainfall below 250 mm), runoff farming appears to be quite effective in terms of water harvesting. Even if the yields are not considered impressive, runoff farming does offer a much greater yield security.

In sum, it may be said that through its control of both water and nutrient availability, the *teras* system is as viable and as sustainable as rainfed farming can get in this harsh environment. No use is made of either petrochemical fertilizer or explicit manuring, and it is very likely that neither would be able to offer both a viable and sustainable alternative to the indigenous system of water and fertility management.

Conclusion

For some two millennia the Beja have been engaged in various forms of runoff farming in the harsh and dynamic environment of north-eastern Sudan. While cultivation probably always remained second in cultural value to livestock production, the Beja have acquired considerable skills in dryland farming. Skills that continue to be further developed by the present-day Hadendoa and Beni Amer of Kassala's Border Area.

The Hadendoa, especially, suffered considerably when during the colonial era the traditional desert trade was supplanted by modern means of transportation and large-scale irrigation schemes occupied the best-watered soils. In effect, they were driven to increase cultivation in the more marginal areas. Presently, the Hadendoa and Beni Amer of the Border Area engage in three forms of indigenous runoff farming on these arid piedmont plains: wild flooding, water spreading, and *teras* water harvesting.

In adapting to the low and irregular rainfall and continually shifting wadi courses of this dynamic environment, individual farmers have developed numerous variations on these runoff farming systems. The knowledge and creativity of the farmers is especially apparent in the complicated and relatively labour-intensive *teras* system. In the farmlands of the village of Ilat Ayot not a single *teras* looks the same. They vary in the size of the field and in shape, height and orientation of the bunds, each *teras* uniquely adapted to the hydrology and soils of a particular site. And, as wadi courses and discharges change, they are re-adapted through continued experimentation.

In coping with (and exploiting) environmental dynamics and change, adaptation and experimentation go hand in hand. The success of the indigenous runoff farming systems does not depend solely on the reproduction of knowledge collected over the centuries. In the dynamic and continually changing Border Area such knowledge would be of only rudimentary value. It needs to be supplemented and continually updated through experimentation, i.e., through the production of knowledge. Knowledge of the environment, of *teras* and wadi hydraulics, and of crop cultivation. While farmers exchange ideas, and some farmers are more enterprising than others, the sheer dynamics of the environment force each individual to produce his own knowledge as it pertains to the particular site he is cultivating. Ultimately, it is this individual experimentation that determines how successful a farmer will be in obtaining a sufficient and sustainable yield to feed his family.

While the environmental dynamics of the Border Area are in some ways perhaps extreme, they are not unlike those in many other (semi-)arid parts of the world. Therefore, the observation that experimentation is crucial in (re-)adaptation of agriculture to changes in the natural and human environments has a more universal validity. Johnson (1972) noted it, but so did, for example,

Richards (1985), and Rhoades and Bebbington (1988), to name but a few. Their works clearly show that successful adaptation has its roots in experimentation, which makes it sad to note that the 'adaptation' of indigenous agriculture somehow has become much more widely accepted than has the active 'experimentation' of indigenous farmers. In fact, in any healthy agricultural system adaptation and experimentation go hand-in-hand in coping with ecological and socio-economical change. Failure to recognize the importance and the widespread existence of indigenous experimentation has led to the current situation where much of the development interventions in third world agriculture are either based on the assumption that indigenous agriculture is backwards, or that it was well adapted, but to no longer existing conditions, and thus in either case needs to be replaced by modern practices and technologies (Niemeijer, 1996).

To take the issue one step further, recognition of the importance of experimentation has two serious implications for the transfer of technologies. First, if indigenous farmers have means to cope with change, support of such indigenous experimentation might prove more effective in dealing with local problems than the transfer of alien technology. Indeed, one might think of the transfer of experimentation techniques and models rather than of technology. This would allow indigenous farmers to expand their ways of experimentation and thus help them solve their own problems. Second, if knowledge transfer does take place, great care should be taken to transfer the general idea rather than a complete package. For local farmers it is much easier to experiment with an idea or concept and to incorporate that into their knowledge system and adapt it to the local environment and their agricultural practices, rather than having to deal with an ill-adapted (because it comes from elsewhere) complete package containing relevant and irrelevant components that take time to separate (if the development agency would in fact allow that to happen). In other words, ideas behind a technique should be transferred rather than replicates of that technique. Transfer of a combination of ideas and experimentation methodologies might prove a very rewarding alternative to the kind of technology transfer that is currently taking place. Scientific experiments may be very important in the development of new technologies and the adaptation of existing technologies, but they can never replace individual experimentation by indigenous farmers, who in the end will always need to rely on the capacity to produce their own knowledge.

Acknowledgments

The author wishes to express gratitude to, among others, the following organizations and individuals: Kassala Department of Soil Conservation, Land Use and Water Programming; Kassala Area Development Activities; Johan Berkhout; Anita van Dam; Ton Dietz; Johan van Dijk; Hashim Mohamed El Hassan; John Houtkamp; Valentina Mazzucato; Rudo Niemeijer; Jan Sevink; Leo Stroosnijder.

5. The indigenization of exotic inputs by small-scale farmers on the Jos Plateau, Nigeria

KEVIN D. PHILLIPS-HOWARD

Abstract

THIS CHAPTER DESCRIBES the way in which exotic inputs are selected, tested and modified to suit local socioeconomic and biophysical conditions. Particular attention is paid to experimentation with, and the incorporation of, inorganic fertilizers, exotic crop varieties and pump-engines. These are analysed in the context of the rapid development of dry-season farming as affected by market forces, structural adjustment and changes in irrigation technology.

Introduction

The concept of indigenization refers to the incorporation by a people of externally derived technology and other resources into their culture. Before elaborating this concept, clarification of the terms 'indigenous' and 'indigenous knowledge' is needed.

In the African context, it has been argued that something is indigenous if 'it is an authentic expression or outcome of Africa's history, social evolution or culture' (Ake, 1987, p. 6). This is largely consistent with the contemporary dynamic view of indigenous knowledge as that developed by indigenous people, or any other defined community, as opposed to knowledge generated by the international 'Western' knowledge system (IDRC, 1993). However, the dichotomy between indigenous and Western knowledge is somewhat misleading because few indigenous systems have developed entirely without absorption or adaptation of knowledge from external sources, including the Western one. In many instances it may be impossible to determine whether, or to what extent, particular knowledge was developed from within. Moreover, Ake's definition of indigenous would not exclude borrowings or influences from outside (Chambers, 1983; Atteh, 1992). These may include externally derived equipment, materials, ideas and techniques which can be locally reinterpreted and transformed. Once people regard such incorporations as legitimate elements of their own culture then, it can be argued, they have become indigenous. Such elements may also be indigenized through modification, including adaptation of form, function, purpose, technique and use. Ultimately, for knowledge to be indigenous it must be felt by local people to 'belong' to them. Such a sense of ownership may be brought about through the very processes of incorporation.

The indigenization of exotic inputs is viewed here as a complex set of processes through which exotic technologies are incorporated by local people into their own knowledge and production systems. In the present context, this involves the perception, selection, acquisition, testing, modification and acceptance of technology, including technical knowledge, in response to local demand and prevailing socioeconomic and environmental circumstances. Hence, indigenization is seen as the active obverse of the passive 'adoption' process conceived in the transfer-of-technology paradigm of development. It is a continuous means by which people satisfy their needs and attain their wants and aspirations in accordance with their changing resource endowment.

In this chapter, particular attention is paid to the incorporation of inorganic fertilizers, exotic seeds and pump-engines by small-scale farmers on the Jos Plateau, Nigeria.

Environment and irrigation on the Jos Plateau

The Jos Plateau comprises an 8600 km^2 area of central Nigeria averaging 1200 m above sea level. Monthly mean temperatures (20 to 24°C) are low for the latitude (10°N), while rainfall (1000 to 1500 mm p. a.) is comparatively high. The landscape comprises an undulating plain of heavily weathered basement complex rocks into which granites and basalts intruded to form rocky hills (Morgan, 1979).

The plain is continually cultivable where water is available, but its deep, sandy soils are poor and susceptible to erosion. The more fertile flood terrace (*fadama*) soils are also cultivable in the dry season (October to April). The shallow, sandy lithosols of the rocky outcrops have been used for terraced crop production as well as grazing and forestry. Some of the basaltic soils are highly fertile and productive.

The natural Guinea Savannah vegetation has been removed from both the plain and many of the hills. The present non-cultivated vegetation on most of the plain, comprises complexes of grass, shrub and woodland. There are 'economic trees' on the farmland and plantations, mainly of Eucalyptus. Whereas about 53 per cent of the Jos Plateau is actively cultivated, the non-cultivated areas comprise: woodland, 37 per cent; shrub/grassland, 6 per cent; shrubland, 4 per cent.

Throughout the Jos Plateau, rapid expansion has occurred in irrigated dry-season farming over the past decade. This spontaneous process was to some extent encouraged by Nigeria's Structural Adjustment Programme (SAP), which apparently stimulated agricultural output (Phillips, 1990) – despite the increased costs of production and inaccessibility of certain inputs. Irrigated agriculture spread rapidly in the 1980s in response to the growing demand for vegetables and the availability of labour, mine-pond water, mechanical pumps and fertilizers (Adepetu, 1985; Morgan, 1985). This change involved increasingly intense, productive and profitable dry-season farming with more educated and ethnically diverse individuals, and expansion beyond the traditional *fadama* areas. In some places it merged with rainy-season farming to form a highly profitable round-the-year system of market-oriented agriculture (Phillips-Howard *et al.*, 1990). The process was further encouraged by the Jos International Breweries outgrower scheme involving irrigated cultivation of barley, mainly on former mineland (Porter and Phillips-Howard, 1993).

This chapter is concerned with a group of farmers who live in the village of Yelwa and farm along the Delimi River, to the north of Jos.

The farmers' problems

Farmers on the Jos Plateau have to deal with numerous problems in their struggle to produce food and crops for sale. The most common of these are shortages of inputs, cash and labour. The major problems identified by 31 of the Yelwa farmers in a 1990 survey are shown in Table 5.1. 'Unavailability of chemical fertilizer' was articulated most frequently. This problem included

Table 5.1. Major problems identified by the farmers

Problem	*No. of farmers*	*%*
Unavailability of chemical fertilizer	26	83.9
Shortage of cash	16	51.6
Lack of pump-engine	14	45.2
Unavailability of seed	11	35.5
Pestilence	9	29.0
Unavailability of transport	5	16.1
Lack of irrigation hose	2	6.5
Feeding of labourers	1	3.2

Note: Many farmers identified more than one major problem (after Phillips-Howard and Kidd, 1990).

both physical scarcity and prohibitive cost. The high cost ($4 to 5 per 50 kg bag at that time) was regarded as more critical, since most farmers said they could scarcely afford to buy it. One farmer complained that he could not afford to buy any inorganic fertilizer. Others stated that their difficulties were lack of stock in the market and inability to buy inorganic fertilizer at the subsidized price (then $1.50 per 50 kg bag) from the Plateau Agricultural Development Programme (PADP). The PADP was allocated only 10 per cent of the 230000 tonnes needed for distribution in 1990. Certainly, this shortage of fertilizer supply to the Jos Plateau was serious, given the high nutrient requirement of the crops grown and the poverty of the soils.

The second most frequent problem, 'shortage of cash' was more fundamental, since many of the other farming problems could be solved by adequate cash flow. This assertion was supported by a *Sarkin Noma* (Hausa for 'chief of farming') who said that without sufficient money, farmers were bound to face serious problems. Many farmers struggled to operate with a cash shortage, spending their income as soon as it was received. This problem was exacerbated by the rising costs of inputs and the declining value of the Naira from around $2.00 in 1980 to $0.13 in 1990. For example, pump-engines quadrupled in price between 1985 and 1990, from around $100 to $400–500. Similarly, the cost of hired migrant labour rose from a nominal amount to $40–60 per dry-season, in addition to the provision of accommodation, food and clothing.

The 'lack of a pump-engine', the third major problem, could impose severe constraints on both farm size and location. Without access to engines farmers in the study area had to use labour-intensive irrigation techniques (e.g., *shadoof*, calabash and bucket), which limited farm size to about 0.1ha and required very close proximity to a water source. For the wealthier farmers the problem was how to acquire more engines to permit farm expansion.

The fourth ranking problem, 'unavailability of seed', seemed to relate more to scarcity than to unaffordability, since the cost per packet was small compared with the quantities needed (e.g., lettuce $0.50, onion $0.40, cabbage, $1.23) and farmers were known to devote considerable time and effort searching for better sources of seed. However, in a few cases seed cost was a problem, especially for the poorest farmers.

'Pestilence', the fifth problem, was recognized by the farmers to be increasing in severity and responsible for a substantial loss of income through crop damage. However, two of the farmers characterized the problem as difficulty in acquiring pesticide, which was comparatively expensive at $3.00 per litre bottle of Gamalin 20 and $0.50 per 100 gm of Permethrin powder.

The sixth ranking problem, 'unavailability of transport', was one either of scarcity (including poor accessibility) or cost. Some farms were relatively far from routes plied by motor vehicles and the cost of transport ($0.10 to $0.20 per basket in 1990) could seem high, especially when crop prices fell, as in January 1989 when a basket of tomatoes could fetch only $0.25.

Lack of 2- or 3-inch irrigation hose was regarded as a major problem by only two of the farmers, though at $5.00 per metre it was comparatively expensive. However, the farmers did sometimes lend hoses and they had alternative means of conveying water over longer distances.

The problem of 'feeding of labourers' was only mentioned once, by a farmer who was responsible for feeding up to 10 seasonal-migrant labourers as a contractual obligation.

With regard to the solution of these problems, outsiders could help bring (Hausa: *kawo*) fertilizer, information, seeds and assistance with transport. Such help is often seen by Plateau farmers as tangible repayment for 'secrets' provided in interviews. This so-called '*kawo* mentality' could partly explain the priority given to supply of inputs, which was seen to be potentially solvable by outsiders. In terms of the factors of production, Table 5.1 shows that only one of the problems related directly to labour and none concerned land. Since the type of farming was labour intensive and land availability was problematic for poorer farmers at least, the non-articulation of land and labour problems could have been because they were not regarded as '*kawo* solvable'.

All of the problems articulated were directly related to capital, as either forms of it or ones that could be solved with cash. This complements the statement of the *Sarkin Noma* on the importance of cash to dry-season farmers.

Following this survey, search and supply experiments (Chambers, 1989) were initiated to help farmers solve their priority problems. The purpose of these was to discover how and from what sources the farmers could gain the inputs they required to further develop and expand their farms on former mine land. The process of searching for sources of inputs and facilitating their supply was carefully documented, as were the problems involved and the ways in which they were resolved. The findings described below are from studies associated with these experiments.

The indigenization of inputs

Fertilizers

Two or three decades ago small-scale farmers on the Jos Plateau relied almost exclusively on 'traditional' fertilizers (Hausa: *Takin gagajiya*) including animal manure, sawdust, compost and ash. Since then various 'modern' fertilizers (*Takin zamani*) have become available. These have been tested, characterized and generally well-accepted by Jos Plateau farmers. Table 5.2 shows how some of the modern fertilizers are regarded in comparison with a few traditional ones. This illustrates accumulated local experience of fertilizer effects on crops in terms of rate of growth, size, yield and specific characteristics (hardness and leaf development). It also shows how indigenous interpretation of modern

Table 5.2. Key informants' characterization of some fertilizers (after Phillips-Howard and Kidd, 1991)

Type	*Characteristics*
TAKIN ZAMANI (Modern fertilizer)	Produces fast growth, more fruits and harder carrots. Expensive and can burn or kill some crops.
- *Kampa* (NPK) 15: 15: 15	Generally preferred, gives rapid results.
20: 20: 20	More powerful, doesn't burn vegetables, 'cold' (*sanyi*).
- CAN (Calcium Ammonium Nitrate	Absorbs water in rainy season. 'Hot' (*zafi*), can burn crops.
- *Supa* (Superphosphate)	Good for preparing new land. 'Cold', doesn't burn crops.
- *Mai gishiri* (Urea)	Very powerful, helps leaf development, best for spinach.
TAKIN GAGAJIYA (Traditional fertilizer)	May not produce high yield unless modern fertilizer is also applied.
- *Tokan bola* (Refuse ash)	Commonly used with modern fertilizer. Softens the soil.
- *Kashin balbela* (Egret manure)	Better than modern fertilizer, very powerful, good for carrots, 'hot', can burn crops.
- *Kashin shanu* (cow manure)	Not preferred, produces weeds.
- *Kashin kaji* (chicken manure)	Powerful and clean, good for onions.
- Compost	Softens soil better than Tokan bola, but not used: unavailable and produces weeds.

fertilizers is reflected in the local names given to them. For example, urea is called *Mai gishiri*, which means 'salt-like one'. Similarly, the compound fertilizer 27: 13: 13 is known as *Wake de shinkafa* ('beans and rice'), because of its appearance. Alternatively, *Kampa* 15: 15: 15 is named either *Dan* ('son of') *Lagos* or *Dan Port Harcourt*, according to the port of arrival in Nigeria, and some *Supa* is called *Dan Kaduna* after the place of its manufacture. This classification by source is important to the farmers because the quality of product is known to differ according to where it comes from. Similarly, traditional fertilizers are classified by their origin, for instance, *Tokan bola* (refuse ash) is literally 'ash of the boiler' and *Kashin shanu* is 'manure of the cow'. Again, this is important because qualitative differences are recognized from one source to another. Table 5.2 further demonstrates that modern fertilizers have been categorized as either *zafi* ('hot') or *sanyi* ('cold') in accordance with their capacity to 'burn' crops.

Such characterization and classification of modern fertilizers suggests that their incorporation into the knowledge system occurred in terms of pre-existing constructs. It also has implications for the use of modern fertilizers, which deviates greatly from standard recommendations. The utility of modern fertilizers is conceived not so much in terms of individual or optimal use as in their potential roles in fertilizer mixtures.

Small farmers on the Jos Plateau combine fertilizers knowing that mixtures are often more effective than individual types used alone. Reasons given by the farmers for mixing fertilizers include softening of the soil, better crop growth and reduced costs. Also, it seems that mixing fertilizers is an adaptation to uncertain availability whereby the risk of dependance on one source is avoided.

Fertilizer combinations are based on various criteria including cost, availability, notions of equivalence and 'heat'. Cash shortage has motivated the farmers to develop low-cost but effective mixtures, such as the combination of refuse ash with NPK 15: 15: 15, which one farmer said can reduce fertilizer costs by up to 60 per cent. When a preferred fertilizer is not available others may be substituted based on their perceived equivalence. For example, CAN plus *Supa* are considered to be equal in strength (*karfe*) to DAP. Similarly, *Kashin shanu*, *K. akuya* and *K. tumaki* (cow, goat and sheep manure) are considered equal and interchangeable in mixtures; they are all 'cold' and suitable for mixing with 'hot' fertilizers such as CAN, Urea and DAP (Diammonium phosphate).

Farmer experimentation has also produced knowledge about crop fertilizer requirements. Although some farmers emphatically stated that all crops have the same requirements, others indicated that requirements differ considerably. Perhaps, as one farmer commented, it is difficult to distinguish particular requirements when fertilizers are broadcast and crops are intergrown. Nevertheless the information on Table 5.3 was corroborated by several informants.

Experimentation is also evident in fertilizer application techniques, again on the basis of the characterization and classification described above. For example, farmers developed alternatives to broadcasting so as to avoid 'burning' crops with 'hot' fertilizers. These included careful placement of the fertilizer, either around individual plants or at the inlet to each basin in which crops are grown.

Table 5.3. Farmers' knowledge of fertilizer requirements of specific crops (after Phillips-Howard and Kidd, 1991)

Crop type	*Fertilizer requirements*
Tomatoes	Need much fertilizer, and prefer modern fertilizer such as NPK mixed with *Tokan bola*; they are easily 'burnt; extra dressings are needed at the time of fruit development; *Kashin balbela* plus NPK will produce large, good-looking fruits.
Leeks	Similarly, need much fertilizer.
Cabbage	Does not need as much fertilizer as the above two; it prefers soft (*laushe*) soil, but can grow on harder soil too.
Lettuce	Does not need additional fertilizer once it has grown; it is easily 'burnt' by fertilizer.
Carrots	Prefer *Kashin balbela* mixed with *Tokan bola*; NPK makes the roots harder; they do not like Urea; excessive fertilizer applications will cause them to rot and damage their leaves; they grow well in soft soil.
Onions	Prefer *Kashin kaji*; they grow well in places where dust rises, e.g., near footpaths or water points; some prefer firmer types of soil.
Sugar Cane	Grows particularly well with applications of *Supa*.
Spinach (*Alaiho*)	Prefers Urea, which helps to produce many leaves; it need plenty of fertilizer, but is also easily 'burnt'.

In the latter case the fertilizer is spread by running water as the basins are irrigated.

Seeds
The Yelwa farmers were extremely eager to acquire what they considered to be 'better' kinds and varieties of seeds. In particular, they wanted to test new types of carrots, cabbages, tomatoes, leeks, beetroot, turnips and onions.

At the request of the Yelwa Farmers Association, vegetable seeds were sourced in Europe and small quantities were provided for the farmers to try out. The trials with these vegetables lasted from January to May, 1991. The seeds were distributed by the Association's chairman and trials were monitored through several rounds of semi-structured interviews.

The farmers' criteria or characteristics for evaluating the vegetable crops are shown in Table 5.4. They seem to include observations on growth performance (germination, growth rate, penetration of soil, leaf drying) and product form (size, shape, hardness, weight) as well as market values (taste, smoothness, size, colour, perishability, etc.). It is not certain that either the list of characteristics or the observations on particular crops are complete. However, assuming that Table 5.4 is broadly indicative of the farmers' assessments, it appears that some characteristics are considered important to a range of crops while others are specific to particular types. Whereas rate of germination, size, yield, shape,

Table 5.4. Characteristics observed by farmers in trials of imported crop varieties

	Crop type				
Crop characteristics	*Carrot*	*Tomato*	*Beetroot*	*Leek*	*Cabbage*
Rate of germination	x	x	x	x	
Strength at germination	x	x	x		
Growth rate	x			x	x
Strength			x	x	
Penetration of soil	x				
Resistance to pests/ diseases	x	x			
Leaf drying		x			
Fruiting		x			
Size	x	x	x		x
Shape	x	x		x	x
Weight					x
Colour	x				
Scarring		x			
Presence of holes		x			
Hardness	x				
Smoothness		x			
Yield		x	x	x	x
Perishability		x			
Views of others	x	x	x		
Taste	x	x			
Marketability	x	x	x	x	x
Sale price	x			x	x

marketability and views of others appear broadly important; others such as leaf drying, penetration of soil, hardness and colour are mentioned only for carrots. Similarly, while occurrence of fruiting, smoothness, perishability, scars and presence of holes are concerns with tomatoes alone, weight is considered solely for cabbages.

All of these characteristics seem closely related to the profitability of crop production. This is consistent with the market-oriented nature of dry-season vegetable production on the Jos Plateau.

Table 5.4 also suggests that the farmers consider more characteristics in assessing some crops than they do in others. For example, 14 characteristics are

Table 5.5. Positive and negative evaluations of crop characteristics

Crop type	*Characteristics evaluated by farmers*	
	Postive	*Negative*
Carrot	Strong, fast germination, Rapid growth. Large fruit. Sweet taste. When mature, leaves do not dry. Resistant to disease. Attractive colour. Preferred by buyers.	Poor ability to penetrate soil.
Tomato	Strong, fast germination. Large fruits. Very sweet in soup. Praised by others.	Scarred surface. Lower yielding. Small fruit with holes. Failure to fruit after transplanting. Low market price. Blotched surface. Tiny seedlings. Poor yield. Easily spoiled.
Beetroot	Strong, fast germination. Healthy looking. Large. High yield. High market price. Preferred type.	
Leek	Fast germination. Strong looking. Rapid growth. High yield. High market price. Praised by others.	
Cabbage		Poor growth. Low weight. Small leaves. Low yield. Narrow shape. Low market price.

mentioned with regard to carrots and tomatoes, but only seven or eight in the cases of beetroot, leeks and cabbages. It is not clear whether such differences are associated with the economic importance of the crops to the farmers. However, tomatoes are the longest grown and most widespread market vegetable on the Plateau, while carrots have recently become very popular there and are commonly grown on sandy soils.

Table 5.5 shows the positive and negative crop evaluations by the participating farmers. The evaluations of the carrot, beetroot and leek seeds are almost entirely positive, while those of the cabbage are totally negative. By contrast, the evaluations of the tomatoes are ambiguous. It may be, as one farmer indicated, that the varied performance of the tomatoes was partly due to differences in management and soil type. However, it could also be associated with differences between the varieties tested.

In general, the crop characteristics were evaluated by comparison with those of locally available varieties. This also applies to water requirements (one farmer observed that the imported varieties needed more water than the 'local' ones). Varieties regarded as much better than those already available included the Boltardy Beetroot and Autumn Mammoth – Argenta Leek. This information was passed on to PADP's seeds officer so that the preferred types could be supplied in bulk to Jos Plateau dry-season farmers. The carrots (Zino and Jasper) were also found to perform well, but further supplies of them were not required by the farmers. By contrast the cabbages (Hispi and Shamrock) and Tomatoes (Red Dawn and Super Marmande) were rejected. In some cases they were uprooted at an early stage to prevent wastage of space.

Pump-engines

Pump-engines are of critical importance on the Jos Plateau, because they enable small-scale farmers to expand agriculture, spatially and temporally. Among the Yelwa farmers, the proportion using engines increased from 43 per cent in 1982 to 84 per cent in 1990 (Phillips-Howard *et al.*, 1990). According to them, a pump-engine (together with sufficient cash to buy other inputs) is a prerequisite for more productive and profitable dry-season farming. The farmers know that a pump-engine can irrigate at least ten times the area of the traditional *shadoof* (which is limited to about 0.1 ha), but at substantially greater cost.

There are four main strategies for increased production involving pump-engines.

- The cultivated area in the dry season can be expanded by extending irrigation systems; increasing water supplies, or making more efficient use of existing water supplies.
- The effective growing season can be extended by: supplementary irrigation during 'gaps' in the rains; early establishment of crops using nurseries; growing drought-tolerant crops when there is a shortage of water, and growing flood-tolerant crops in areas subject to flooding during the rainy season.
- The available land and water, and the effective growing season, can be more fully utilized through: production of higher-yielding crops and varieties; greater use of manures and fertilizers; greater use of herbicides and pesticides; maintenance of continuous crop production; relay cropping, intercropping and multiple cropping.
- The labour input per unit area can be increased to improve land preparation

and soil cultivation, fertilizer and pesticide application, weeding, harvesting and water management.

The Yelwa farmers seem to apply each of these strategies. In so doing they incorporate the engines into their system of small-basin irrigation using small earthen canals, which are known to save much time and labour (Cottingham, 1988).

For example, according to their economic circumstances, the farmers have devised various ways to expand their irrigated land. These include: pumping up to high points, or earth and stone ramps (Hausa: *Dokki*), from which the water then flows by gravity; relay-pumping, where water is pumped into a distant pit from where it is pumped again to higher land; excavation of channels, removal of mounds and construction of hose or pipe systems to reach otherwise inaccessible areas; and simultaneous use of two or more engines.

The incorporation of pump-engines has necessitated various adjustments, including greater input of labour and faster work. In particular, farmers have

Table 5.6. Farmers' knowledge of water requirements

Characteristic	*Water requirement*
Crop type	As seedlings, all crops have the same water requirements.
	Leafy crops (e.g. cabbage, *alaiho*, lettuce, parsley and celery) need daily watering to develop well, though cabbage needs less at maturity, otherwise it will 'burst'.
	Crops that produce leaves and 'fruit' (e.g. carrots, onions, leeks, tomatoes, and beetroot) should be watered only at 3–4 day intervals, otherwise the fruit may rot.
Land type	Land close to the river is 'cold' (Hausa: *Sanyi*) and requires less frequent watering (4-day intervals are sufficient) than land which is distant from the river.
Soil type	*Rai rai* (very soft soil) absorbs water more quickly than *Tabo* (sticky black, red or white soil which holds water and can be irrigated at 3–4 day intervals), but the latter needs to be given more water because it absorbs it slowly.
	Palele (whitish, infertile, fine and shiny, cannot retain water, needs frequent irrigation) is not good, it absorbs water slowly.
Weather type	In the late rainy (*Kaka*) season (when nurseries are established), little water is needed, dew is present.
	During the cold *Harmattan* period of the dry season, little water is needed. Irrigation at 2–5 day intervals is sufficient.
	During the hot dry-season (*Rani*), water requirements are very high; daily irrigation is necessary.
Basin age	Old basins have more plant cover and harder soils; they absorb water more slowly than new basins, which have soils that are 'soft' (Hausa: *Laushe*).

had to learn how to manage the comparatively rapid flow of water so as to avoid crop damage and basin overflow. This has involved learning how to adjust the flow rate of the engine, and increased knowledge of variations in water requirements. For example, measurements of irrigation time of different standard sized basins indicates adjustment of flow rate according to soil type. On average, water was allowed to flow for 31 seconds (range: 8 to 49 seconds) into sandy basins close to the river, whereas the average flow time on a distant clay soil was 132 seconds per basin (range: 63 to 197 seconds).

Table 5.6 shows that the farmers know how water requirements differ by crop, types of land, soil, weather and age of basin. It is evident that the knowledge of water requirements is based largely on the farmers' classifications of crops, soils and seasons.

Most (57 per cent) of the farmers interviewed regarded 'saving to buy' as the solution to the lack of a pump-engine. However, a search and supply experiment revealed that poorer farmers are unable to save enough even to buy a highly subsidized engine (sold by PADP in 1992 for $225 equivalent as opposed to the market price of $300). For social reasons, shared ownership was not seen as an option for the poorer farmers. Alternatively, hiring and borrowing of engines were seen as possible solutions.

Conclusion

Interest has recently been expressed in a new paradigm of sustainable development based on the role of indigenous knowledge systems in agricultural improvement (e.g., Titilola, 1992). One approach toward the achievement of such a locally based sustainable development process, suggested by Slikkerveer (1993), involves the transformation of technology to meet local needs and expectations. This chapter described how one group of small-scale farmers engage in this process through the indigenization of exotic inputs.

The evidence presented here indicates that small-scale African farmers are well able to 'transform' such diverse technologies as fertilizers, seeds and pump-engines as they incorporate them into their own knowledge and production systems. This indigenization process apparently involves learning about the characteristics, value and 'malleability' (flexibility of use) of these inputs and the extent to which they could be fitted to local needs and purposes. The degree and rapidity of such learning may have varied according to the familiarity, simplicity/complexity and accessibility of the technology as well as the farmers' aptitudes and motivation to incorporate them. In the case of the much-needed inorganic fertilizers, learning and incorporation seemed to be relatively thorough as a result of several years of testing and experimentation. By contrast, pump-engines were incompletely incorporated because of their inaccessibility to the poorer farmers, and possibly their complex management requirements. Vegetable seeds showed potential for rapid incorporation as farmers were able to test and assess them quickly.

Limitations to the indigenization process seemed to include contextual factors (e.g., shortages of land, labour and capital, unacceptable risk, inadequate time, market conditions, weather variations) and limits of individual technologies (e.g., the unsuitability of some hybrids for seed production by farmers, pump-engine capacities, the susceptibility of tomatoes to rot, and individual fertilizer characteristics). Understanding of these limitations apparently enables farmers to incorporate technologies adaptively by matching them with their needs and purposes.

Ultimately, indigenization is a means by which local people capture technology for their own sustainable development; it is an important process that warrants a major research effort.

6. Farmer management of rootcrop genetic diversity in Southern Philippines

GORDON PRAIN and MARICEL PINIERO

Abstract

THIS CHAPTER DISCUSSES some of the results from a long-term study of farmer maintenance of rootcrop genetic diversity in the hill areas of northern Mindanao, southern Philippines. The action research component of the project, involving the exploration of alternative institutional arrangements to support existing conservation practices, is briefly described. The aim of collaborative work with several farmer groups was to assess the feasibility of building a more sustainable, public culture of conservation through scaling up existing household-level maintenance to various forms of communal action. The bulk of this chapter reports the results of monitoring the maintenance of diversity over several seasons by the women participants in the project and also analyses the dynamics in terms of the effects on the diversity of the local system and in terms of individual rationale. High levels of diversity for all rootcrops in the first few seasons can be ascribed to the effects of the project intervention. Subsequently there was a slight decline towards a relatively stable level of diversity for sweet potato, taro and cassava, but a disappearance of the less commonly grown yam and yautia. An analysis of the practice and rationale of individual farmers indicated that within the cultivars being grown, they maintain a set of 'dominant cultivars' which cover a wide genetic range and satisfy specific needs. Several of these cultivars are maintained by all farmers from season to season. The study revealed that over the course of the seasons, all farmers increased the average number of these cultivars being grown, indicating a hightened sensitivity to the benefits of diversity. Evaluation of novel cultivars was based first on survivability under 'hands-off' management, and then on the criteria satisfied by the dominant cultivars.

Introduction

This paper discusses some of the results from several years' collaborative work with farmers in Bukidnon province, on the island of Mindanao, southern Philippines, which aimed to understand the dynamics of on-farm biodiversity management and assess the opportunities for long-term *in situ* conservation of rootcrop varietal diversity.

The work followed on from earlier documentation studies of indigenous beliefs and practices associated with sweet potato agriculture and, in particular, knowledge about and use of sweet potato varieties (Nazarea-Sandoval, 1991; 1994a). The results of that study illuminated several important aspects of farmer maintenance and use of diversity.

- The two most common criteria for discriminating between different varieties are morphology and culinary characteristics.
- Preference for a particular characteristic of a variety is sometimes 'fuzzy', in that the positive aspect is tempered with negative qualities.
- Varieties are evaluated as a totality, not for superior performance along only

one dimension; there is thus a balancing of the various characteristics, in the overall evaluation.

- Appreciation for the total 'gestalt' of particular varieties leads farmers to regret the loss of favoured ancient cultivars and to wish for access to planting material to recover them.
- Though farmers may cultivate one or two widely selected varieties on most of their planted area because of their good marketability, they also maintain other varieties on a smaller scale for home consumption.
- Farmers like to keep several varieties to give themselves a choice – 'like changing clothes' – and also to satisfy particular preferences within the family.
- An increase in commercial agriculture was one probable cause of the decline in sweet potato varieties from 22 to 13 in one of the case villages from 1980 to 1990.

These results indicate that local maintenance of cultivars exists alongside the contemporary use of agricultural technology and amidst large-scale socio-economic change, and is based on the diverse characteristics of those cultivars. Nevertheless, economic change also appears to impinge on both the capacity and the desire of rural households to manage large amounts of crop diversity. In other words, it is not clear whether the management of diversity is a stable situation, or whether it is part of a long-term trend towards the final disappearance of many local cultivars in such systems. The study suggests on the one hand the *feasibility* of local maintenance of crop genetic diversity as a modern conservation strategy, but casts doubt on its sustainability. On the other hand, it highlights the *necessity* of understanding more clearly the processes through which farmers evaluate and select novel cultivars and the extent to which cultivars move out of particular local systems.

Two kinds of follow-up on this study were identified. There was a need for action research to explore whether certain types of communal action might help to strengthen and publicly legitimize what are normally private maintenance practices in the face of strong socioeconomic pressures on households that appeared to lead to cultivar losses. There was also a requirement concurrently to study and monitor the processes of cultivar management over time. It was felt that both of these types of research activities could contribute to the on-going debate on farmer management of biodiversity and the potential of *in situ* conservation of crop genetic diversity (e.g., Altieri and Merrick, 1988; Brush 1991;1992;1993; IPGRI, 1993: 38ff). The following research objectives were therefore set.

- Initiate linkages between scientists and farmers in the collection, and *in situ* maintenance of different land-races and wild relatives of locally important rootcrops.
- Document associated indigenous knowledge and maintenance practices over time.
- Develop and test methodologies for communal conservation.
- Evaluate the social and biophysical benefits and problems of different institutional arrangements for *in situ* or 'communal-based' rootcrop conservation.
- Consider the implications of *in situ* conservation for national crop genetic conservation policy.

This chapter limits itself primarily to addressing the first two objectives of the study, especially the process of cultivar maintenance over several seasons. For a fuller discussion of communal approaches to managing local crop genetic diversity, see Prain and Piniero 1998.

Research background

Why rootcrops?

In the prehistoric and historical development of Asian agriculture, there have been two major systems: swidden or slash-and-burn cultivation and irrigated rice (Geertz, 1963; Yen, 1974: 276–78). Yen, summarizing considerable ethnological and archeological data, argues that the establishment of swidden systems throughout peninsular Asia probably preceded the introduction of grain agriculture from the Asian mainland, which occurred around 4000 BC. Rootcrops were the dominant swidden crops during the initial expansion, and have remained core providers of energy food for the household and forage and starchy feed for livestock, especially pigs. The introduction of grain agriculture did, however, lead to the dominance of upland rice in some swidden systems of South-east Asia. Among the rootcrops, taros and yams probably spread very early from the Asian mainland and, in the case of *Colacasia esculenta*, perhaps from eastern Indonesia. In the sixteenth century and maybe even earlier (Yen, 1974) sweet potato and cassava were added, superseding the Asian cultigens in many areas.

The rootcrop complex thus has great antiquity within the region, and has been of major importance as a food source and for supporting the ritually and politically important pig culture. Today, farmers and homegardeners habitually plant together the different types of rootcrops, matching their different adaptive characteristics to variations in the planted area and variable needs for food and feed over the agricultural cycle. It seems wholly appropriate, therefore, to focus a conservation effort on this complex.

Furthermore rootcrops continue to be important sources of both food and income. To underline the importance of rootcrops and other 'secondary' crops, which are frequently associated with marginal, upland environments, Gelia Castillo refers to their 'primary functions' (Castillo, 1995). These functions go beyond the crude division between 'staple' or 'subsistence' and 'cash crop' or 'commercial'. They cover emergency circumstances, seasonal flexibility, nutritional supplements, food diversity and processing functions (Prain, 1995). Genetic diversity is a key support to the diversity of functions since different functions are often best served by different genotypes.

The area of study

Mindanao is the southern-most of the larger Philippine islands and enjoys a more uniform and wetter climate than northern parts. The 'dry' season, from December to March or April is in reality a period of lower rainfall. Crops are planted during this period, though with higher risk from periodic drought. It is easier to maintain sweet potato vines as planting materials through the year than in the northern Philippines – a factor of great importance for a rootcrop conservation garden – but there is always a danger that they may dry up during the low rainfall period.

Bukidnon, the province where this research took place is located in north central Mindanao, an undulating plateau of grasslands cross-cut with deep river valleys which radiate out like spokes of a wheel from the central volcano,

Mount Kitanglad. The plateau is bounded by a once densely forested mountain chain to the east and lower hills along the western border.

Up until the early part of the twentieth century, the plateau was very sparsely populated by Maranao Muslim settlers in the south and Bukidnons and Manobos in the north (Edgerton, 1982: 364). Populations were located mainly in swidden settlements along rivers or in the hills, with few settlements on the grasslands themselves. During the twentieth century there has been a slow cyclical shift both in location of settlements and in agricultural activity. With the introduction of plough agriculture accompanied by a politically motivated drive to relocate populations on the grasslands so as to better control them, the grasslands saw the growth of towns and an increasingly vigorous agriculture, with increased areas of corn, rice and sweet potato.

A reverse shift of population began after the Second World War with the advent of logging operations. First as labourers for the loggers and then as farmers of the deforested land, migrants from the grasslands and from further afield adapted plough agriculture to the hilly areas, especially for corn, rootcrops and more recently, for vegetables. Another production system which is very common throughout the province and which normally supports a large amount of biodiversity, is the homegarden. A study of the role of homegardens in the same province showed that as many as 60 distinct species of fruit trees, food crops and ornamentals are planted, often with several varieties of particular species (Boncodin and Prain, 1997).

In these different production systems, rootcrops have played an important role as supplementary staples, and until quite recently there was a steady increase in the number of cultivars grown because migrants brought with them their favourite cultivars to plant in their new farms. However, the growth in diversity may have peaked around the early 1980s, and decline set in at the same time as agro-industrial and other markets opened up and other of the pressures mentioned earlier have accelerated.

In the areas in which this project has been active, the ethnic composition consists of a mixture of indigenous tribal groups, especially Talaandig and Manobo, and *dumagat*, the 'people from the coast', not just from the north coastal region of Mindanao but especially from the smaller islands further north, such as Bohol and Negros.

Methods

The methods used in this project have drawn from the menu of tools and approaches collectively known as participatory rural appraisal or PRA (Chambers, 1992). These have been particularly important for the initial understanding of local agriculture and associated farmer conservation practices. The study also drew on two sets of methods specially developed for working with local knowledge of crops and crop diversity. Collection of crop diversity and indigenous knowledge (Prain *et al.*, 1995) permits relatively rapid documentation of the agro-ecological context of collected crop samples and the local knowledge about those samples. Memory banking (Nazarea-Sandoval, 1994b) includes a set of procedures and techniques for carrying out a systematic, long-term documentation of both the genetic diversity of an area and the associated knowledge base – the memories – held by local people and which is evolutionarily entwined with the genetic make-up of particular cultivars. The latter is particularly well adapted for the long-term commitment to a particular site which is implied in *in situ* conservation.

Towards a 'public culture of conservation'

Although the exploration of ways to develop a public culture of conservation among farmers as a means to strengthen local agriculture and local choices was an important part of this work, only a brief description is given here of that aspect.

Long-term perspectives, linking with existing experience

The action research component of this project explored the possibility of strengthening local maintenance of diversity through 'scaling up' individual management of cultivars to a 'communal', more public type of conservation. Four different kinds of farmer groupings were identified as partners in this work. Selection of the groups was based on hypotheses emerging from the earlier study about likely factors influencing both conservation behaviour and communal dynamics: gender, formality/informality and ethnicity were the three main variables identified. The research was transparently presented as 'an experiment' in joint conservation of rootcrop diversity. The collection of local diversity, the design of 'conservation gardens' and maintenance activities were all determined by the groups themselves. Some simple incentives were used to support the existence of the groups and the sites during the first year, but researchers did not intervene directly in maintenance practices; rather changes in the conservation sites were monitored.

The preliminary negotiations over the establishment of the conservation activity and the early stages of implementation involved extensive discussions of the long-term characteristics of the project. Conversations focused on the potential contribution of the different 'classes' of local rootcrop cultivars to long-term agricultural security and adaptability. Concrete examples were used for discussion: the fact that the disappearance of local rice varieties had meant loss of taste quality; the fact that the unpredictable environment, such as the eruption of Mount Pinatubo, can lead to new agricultural conditions, in which some varieties prosper and others produce nothing. These discussions brought alive the contribution of diversity to quality of life now and in the future, and also to future adaptation to unpredictable conditions.

The response to these discussions was variable, both between individuals and over time. A real difficulty, of course, is that the notion of 'crop genetic conservation' is an abstract ideal of our making and tends to be conceived of in 'global environment' and 'global interest' terms. In other words, the scientist, policy-maker or other 'specialist' adopts the perspective of 'looking from the outside in', of perceiving the environment, or the crop genetic resources component of the environment as something outside of ourselves (cf. Ingold, 1993). This perspective does not easily connect with the local, concrete concerns and outward-looking perspectives of farmers. What farmers do in their gardens and fields is try to ensure a diversity of food sources, income sources and also, sometimes, sources of intellectual and aesthetic enjoyment. This was clear enough when we collected samples of the wide range of homegarden species for discussion and as an entry point for discussing intraspecific diversity. Availability of food through the year, availability of different tastes, the ease of daily collection, the attractiveness of ornamentals and even of difference itself were all mentioned to explain the choice of 'diversity'. The situation with intraspecific diversity is analogous. It can be seen that this thinking is concrete, largely use-oriented and concerned both with spatial diversity (adapting crops and cultivars to the limited area and agro-ecological heterogeneity of the house and farm) and with temporal diversity in the medium rather than long term (food and income flow from crops and cultivars through the agricultural year). Nevertheless, some farmers did identify a long-term interest in maintaining both species and varietal diversity in relation to their children's and grandchildren's food and income sources and choices and this is one entry point for building a more public culture of conservation.

Livelihood benefits from conservation

There are two main kinds of livelihood benefits that farmer partners felt might be derived from this work: those coming directly from the research team as 'incentives' and some kind of social or economic benefit resulting from the conservation activity itself. The incentives, which were mainly provided during the first year, aimed to contribute to group-strengthening and public legitimacy. For example, organizing cross visits to other villages, provision of certificates of membership, offering small 'diversity prizes', helping to provide noticeboards and fencing, and so on.

The project gave too little attention early on to shorter-term social and economic benefits that could be linked to 'public' conservation efforts, despite the fact that the idea of livelihood was incorporated into the name of one of the farmer groups[1]. Their particular emphasis on livelihood in the here and now rather than on possible future benefits to the agricultural system probably accounted for the loss of interest and termination of activities. In the longest-running farmer group, where awareness certainly grew of the long-term benefits for the community of maintaining diversity in their crops, there were also some 'spill-over' social benefits, such as heightened recognition of the roles and activities of women, including their contribution to the environment. In one case this contributed to the election of a leading member of one women's group to the local council. In another, the group's activities were used as a model by the municipal authorities. These phenomena are the beginnings of a legitimation process which the project sought to explore.

One of the groups, a formal organization of rural women, developed a more direct relationship between conservation and family welfare activities through a focus on home gardening. The women's group had already had experience of communal home gardening projects aimed at improving household nutrition and it was therefore decided to combine conservation of local rootcrop diversity with the cultivation of legumes and vegetables to add to the supply of both food and income from the activity. This and other opportunities for linking livelihood considerations with more stable levels of diversity maintenance will be discussed in the conclusions.

Farmer maintenance of rootcrop cultivars

A key part of the action research designed to explore the potential for communally based, publicly recognized conservation efforts is the monitoring of maintenance activities to assess their effect on local genetic diversity.

Diversity in local systems

The assessment of changes in rootcrop cultivar portfolios, and thus of rootcrop genetic diversity,[2] can be considered in terms of the actions of individual farmers and also in terms of the totality of actions in relation to the local 'pool' of rootcrop germplasm. The impact of maintenance actions on the local system is what we are ultimately concerned about when considering the viability of *in situ* conservation as an option for crop genetic-resources management, but we also need to understand how and why particular changes in local genetic diversity occur. This section looks at overall change in one location where women semi-independently managed conservation plots.

In the village of Maambong, an informal grouping of women – the 'industrious mothers' – determined to establish a 'conservation garden' following the lengthy involvement of several members with a research activity documenting indigenous knowledge of rootcrop agriculture, and particularly of rootcrop genetic diversity. The garden, located on land donated by the eldest member of the group – 'so that I may be remembered' – was modelled by the women on the idea of individual household gardens, so that each woman cultivated their own plot containing her own cultivars and there was therefore considerable duplication of some cultivars across the different plots. After some initial experimentation, the women adopted a rotation scheme with peanuts for replanting their collections, and this allowed more than six seasons of maintenance in this location. Monitoring data was analysed for five replantings.

The dominant and most diverse crop in the conservation garden was sweet potato, which was also differentiated from other rootcrops by its shorter growing period. There was a sharp rise in sweet potato diversity in the second planting from 11 to 19 cultivars, with one loss (Table 6.1). This increase can be explained by several factors. Considerable interest and enthusiasm built up among the group, and even among other villagers during the first planting and as the crop matured. The women regularly met together in the garden for weeding activities; they received frequent visits from the researchers for meetings and discussions; there was a distribution of certificates of membership in the group; and finally, prizes were awarded at the first harvest for the greatest number of rootcrop cultivars maintained and successfully replanted. This led the women to scour the locality for additional cultivars.

The spread of enthusiasm also led to five new members joining and they added new cultivars. Two women also withdrew at that stage, one through illness and the distance of the garden from her house, and the other through lack of sufficient family labour.

From the second to the third planting there was an overall decline in total cultivars. The second planting took place in an adjacent plot (part of the early experimentation with garden design mentioned above), which was rather more exposed both to rodent damage and the invasion of weeds. Both these factors influenced the loss of cultivars, especially cultivars held by only one or two women and represented by only one or two plants. Why did the women not weed their plots? A major reason was the increase in commercial tomato cultivation in the area during 1993, which occupied some women as growers and others as labourers. There were other factors also, such as discouragement caused by extensive rat damage and the competing domestic obligations.

The disappearance of eight cultivars at the fourth planting can be explained partly by the withdrawal of several women from involvement, with some loss of rare cultivars being grown by them. In addition, the earlier planting decisions of the women who remained active also had an effect. They had increased the area in their plots devoted to a set of preferred varieties (*klarin, 5-finger, igorot pula, igorot puti, amerikano, kamada*) and reduced the number of plants of 'residual'

Table 6.1. Sweet potato cultivars planted in conservation garden, Maambong

First Planting	*Second Planting*	*Third Planting*	*Fourth Planting*	*Fifth Planting*
klarin	klarin	klarin	klarin	klarin
5–fingers	5–fingers	5–fingers	5–fingers	5–fingers
igorot pula	igorot pula	igorot pula	igorot pula	igorot pula
tapol	tapol	tapol	tapol	tapol
amerikano	amerikano	amerikano	amerikano	amerikano
bilaka	bilaka	bilaka	bilaka	
igorot puti	igorot puti	igorot puti	igorot puti	igorot puti
kamada	kamada	kamada	kamada	kamada
valencia	valencia	valencia	valencia	
kinampay	kinampay	**kinampay**	kinampay	kinampay
kaligatos	kaligatos	kaligatos	kaligatos	**kaligatos**
	tinangkong	tinangkong	tinangkong	tinangkong
	kapitlok	kapitlok	kapitlok	
	kabohol	kabohol	kabohol	
	sil-ipon	sil-ipon	sil-ipon	
	magtuko	magtuko		
	maranding	maranding		
	senorita	senorita		
	imelda	imelda		
	kitam-is	kitam-is		
		lila	lila	
		initlog	initlog	
			PNGL	PNGL
			kawakwak	kawakwak
			P16	P16
			NPSP	PNSP
			salayaw	salayaw
			UPLSP	UPLSP
			kabato	kabato
			turay	turay
Totals 11	19	17	17	13
Additions	+ 9	+ 3	+ 8	+ 1
Losses	– 1	– 5	– 8	– 5

Note: Bold cultivars are additions
Shaded cultivars are losses

cultivars. The increasing demands of the commercial tomato production work (8 of the 12 were either cultivating or labouring in tomato fields) may also partly explain these planting decisions, but it may also reflect a more widespread tendency, which will be discussed below.

A different pattern of conservation is evident in the long-maturing rootcrops. There is a reduction in the diversity of taro, for example, from the original nine cultivars to four cultivars by the fourth planting (Table 6.2). This is partly explained by the initial maintenance plan, which involved replanting the garden in an adjacent area after the sweet potato harvest, requiring members again to search out planting material from their own fields and houseplots, since the original collection was still immature. The later design of the conservation garden in which a separate cycle of replanting is maintained for the longer-

Table 6.2. Long-maturing rootcrop cultivars planted in conservation garden, Maambong

First planting	Second planting	Third planting	Fourth planting	Fifth planting
Taro				
ginabii	*ginabii*	*ginabii*	*ginabii*	*ginabii*
gabi tsina	*gabi tsina*	*gabi tsina*	*gabi tsina*	*gabi tsina*
salayaw	*salayaw*	*salayaw*	*salayaw*	*salayaw*
paagdaga	*paagdaga*	*paagdaga*	*paagdaga*	*paagdaga*
karagwa	*karagwa*			
wild taro	*wild taro*			
taod-taod	*taod-taod*			
dakan	*dakan*	*dakan*	*dakan*	*dakan*
matagpunay	*matagpunay*	*matagpunay*	*matagpunay*	
	kabang	*kabang*	*kabang*	
Total 9	6	7	4	5
Additions	1	1		1
Losses	4		3	
Cassava				
dilaw	*dilaw*	*dilaw*	*dilaw*	*dilaw*
puti	*puti*	*puti*	*puti*	*puti*
hawayan	*hawayan*	*hawayan*	*hawayan*	*hawayon*
kabutho		*kabutho*		
	kalibre	*kalibre*	*kalibre*	
Yam				
kinampay		*kinampay*	*kinampay*	
talamisan		*talamisan*		
matagpunay				
bulakan				
hinal-o				
ubi-tapol				
binotelya	*binotelya*	*binotelya*		
		katuray		
Yautia				
bisol	*bisol*	*bisol*		
kuryoso	*kuryoso*	*kuryoso*		
sudlon				

maturing species of taro, yam and cassava from that of sweet potato has improved the potential for conservation. Nevertheless, the new planting scheme did not lead to the reintroduction of the cultivars that disappeared after the first planting. The pattern resembles sweet potato, in that the most common cultivars, planted by several members of the group become the stable core of the collection.

Cassava also shows a stable pattern, more so than either yam or yautia, which is probably due to the relatively small number of cassava varieties available in this area and their wide distribution among households. Yam is not a widely grown rootcrop in this part of northern Mindanao, but it is very widely grown on the nearby island of Bohol, from where many of the group members had migrated, and to which they returned for quite regular visits. Most of the yam varieties were introduced from Bohol and the increase at the third planting occurred after the return of the group's elder from a visit there.

In the case of the three major rootcrops, after a variable amount of fluctuation with additions and losses, maintenance stabilized around a core set of cultivars for each species. Why is this? To understand, we need to look at change and stability in diversity maintenance from the point of view of the farmers.

Cultivar maintenance practices

General patterns

The women's management of sweet potato cultivars was characterized by the regular acquisition and divestment of varieties over the last three plantings analysed, after an initial large increase at the second planting (Figure 6.1). The increase resulted from a more intensive search for local cultivars not previously included in the collection, which in turn was stimulated by the build-up of enthusiasm during the first season, the collaborative spirit that developed and the awarding of small prizes for the greatest diversity during the first harvest. There is also an overall increase in the average diversity of individual plots, from just under four cultivars per person planted initially (with a range of from 1 to 8) to a highest average of nearly eight (ranging from 6 to 9) at the fourth planting (Figure 6.2). The generally high number of cultivars maintained by the fourth planting and the high mean additions at that time have a number of explanations. It is partly due to the reduction in number of active participants to a core group who had maintained a higher average number of cultivars from the beginning. At the same time, the group had increased opportunity to add more cultivars through cross-visits to other conservation sites and through requesting and receiving specific types of advanced material from different Philippines breeding programmes.

If we compare the pattern of taro variety maintenance by the women who continued as part of the group (Figure 6.3) with the overall level of genetic diversity over the five seasons seen in Table 6.2, there is considerable difference in the 'shape'. This is primarily because five out of the nine cultivars in the first season were planted by one woman, with the other women planting only one or two cultivars. The subsequent increase in average number of cultivars planted looks quite similar to the management of sweet potato, except that the peak is slightly later. This is because the original taro cultivars had not matured when the new site for the conservation garden was planted, and women did not have the opportunity to use the range of cultivars still growing in the old garden. The third planting took place in the original site and the original plantings of taro were then available as planting material for different members. In other words, though overall genetic diversity declined after the first planting, more of the women made an effort to increase the diversity of their own plots, thus pushing up average taro diversity. Again, the stabilization in average numbers of cultivars over the fourth and fifth plantings indicates common maintenance of a core set of taro cultivars (Figure 6.4).

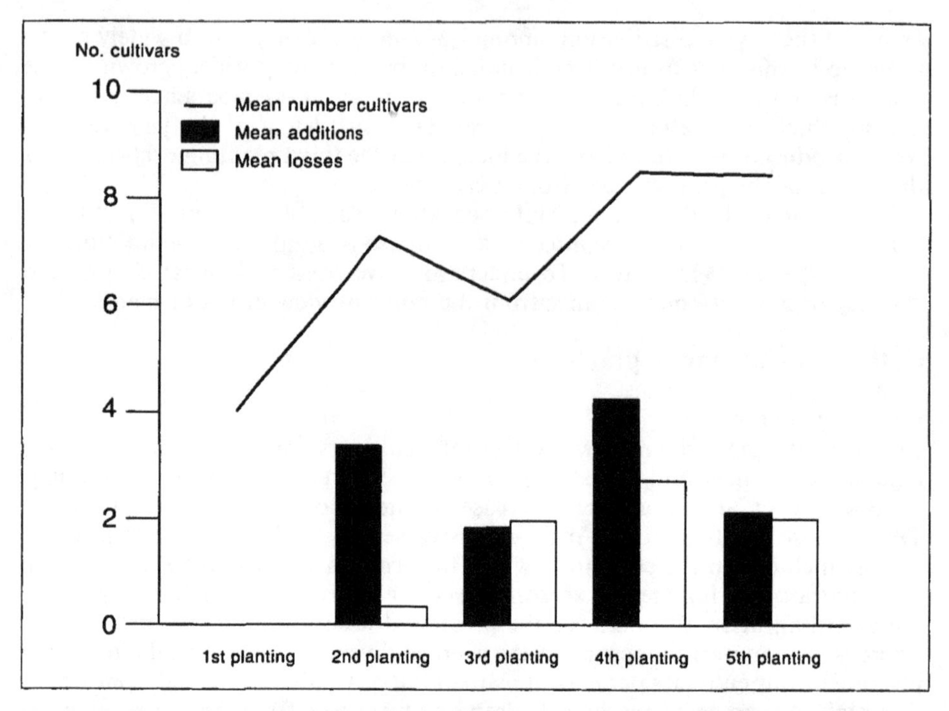

Figure 6.1 Farmers' additions and losses of sweet potato cultivars over five seasons in Maambong conservation garden

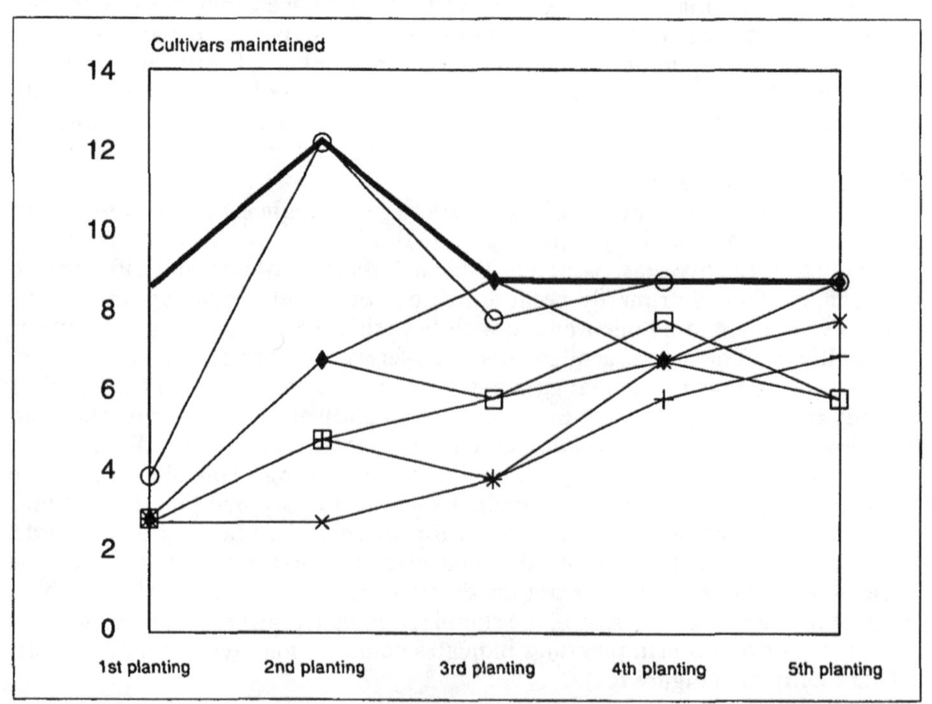

Figure 6.2 Pattern of sweet potato cultivar maintenance over five plantings by seven Maambong women

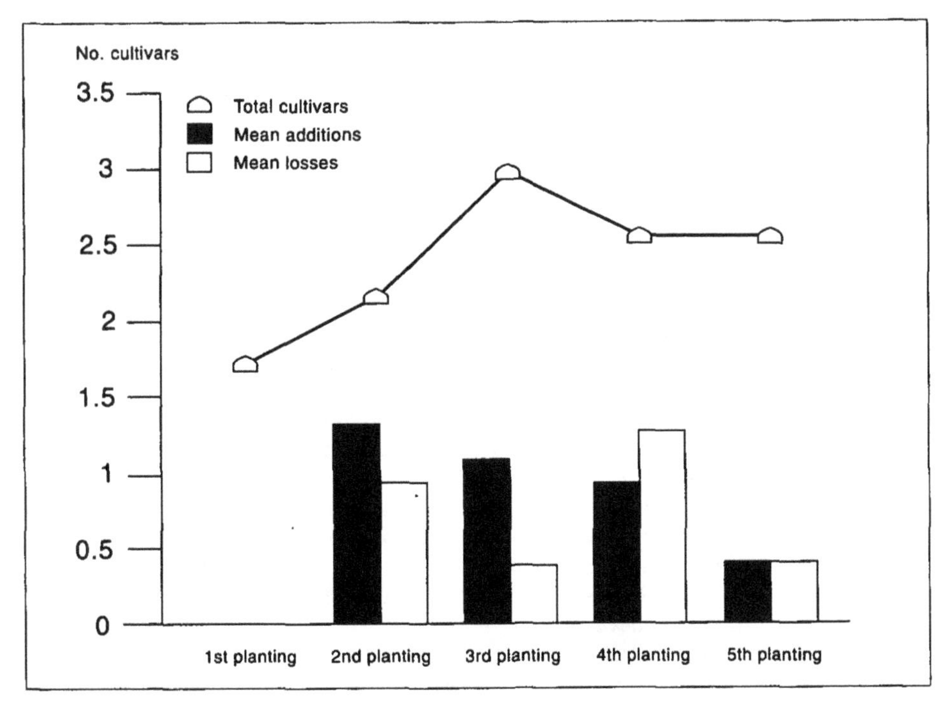

Figure 6.3 Farmers' additions and losses of taro over five seasons in the conservation garden, Maambong

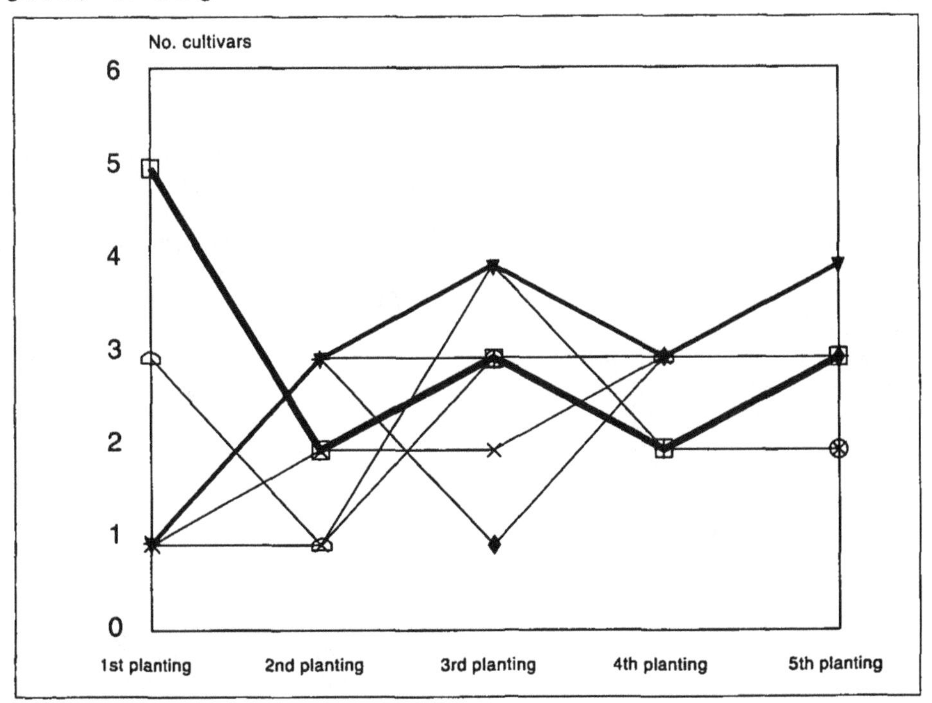

Figure 6.4 Pattern of maintenance of taro cultivars over five plantings by eight Maambong women

Dominant and residual cultivars

Although a number of factors, such as cropping characteristics and the composition of the conservation group, help to explain overall patterns, they do not explain why some cultivars are retained while others disappear from particular women's plots. We have seen that the women tend to converge on a core set of cultivars in both sweet potato and taro. These cultivars are 'dominant' in the sense that they are mostly present at the beginning in at least some of the women's plots, and gradually spread to most plots by later plantings. If they disappear from a plot at one planting, they are likely to be recovered and replanted subsequently. They can be contrasted with other types of cultivars which last only one or two seasons and then disappear.

What is special about the 'dominant' cultivars in Maambong village? Some of the cultivars are grown because they are popular in the fresh market and this applies especially to *Klarin* and *5-fingers* (Table 6.3). These varieties are most commonly found in field production and they tended to occupy the largest area of the conservation garden. However, not all these dominant cultivars are grown for the fresh market and several are grown only in a limited area of the garden. Yet they are consistently present. The most important feature of Table 6.3 is the way the preferred cultivars show quite a wide genetic diversity, as reflected in the range of values of morphological descriptors, and also a wide range of possible uses. Apart from differences of vine, leaf, root and flesh colours, there is also variation in taste according to farmers, and variation in use, from home consumption of roots and vines, to use as forage and sale for fresh or processed uses ('*comote cue*'). This seems to support the idea that genetic diversity of cultivated crops is closely related to the diversity of uses to which they are put and that these multiple uses need to be maintained or expanded in order to ensure the continued genetic diversity at local level (cf. Vega *et al.*, 1998).

A second point is that at the beginning of the conservation garden all the dominant cultivars were present, but individuals were not necessarily cultivating them all. By the fifth planting, as has been observed above, most of the women still involved in the garden were managing most of these dominant cultivars, even if some remained limited to a small area. Our interpretation of this is that preferred types are known and recognized and may have constituted a kind of 'ideal set' in the past, but new constraints and concerns had led some households to retain only one or two types in their own system. The 'conservation experiment' seems to have led to recovery of most of the set, at least in the garden, and over the five seasons.

Some support for this interpretation comes from spontaneous events, uninfluenced by external researchers, that occurred in another conservation garden site, following the abandonment of the experiment by the local partners.

The Dalwangan site was originally set up under the responsibility of a tribal authority with a complex hierarchical management structure. The authority was able to organize the establishment of a single, communal garden, and the collection of a large number of both known and novel rootcrop cultivars from the vicinity. Site organization quickly changed, however, as it became clear that the project would yield few short-term livelihood benefits and was given low priority by the tribal chief. A brother of the chief briefly took over responsibility as the site was located on his own land, but he soon lost interest. With these structural changes, there was a large loss of cultivars between the first and second plantings and shortly after, abandonment of the garden.

One of the local women who had been involved in the day-to-day management

Table 6.3. Morphological and farmer characterization of continuously planted, dominant cultivars in the conservation garden, Maambong

Cultivars	Plant type	Vine colour	Leaf vein colour	Leaf shape	Imm/ mature leaf col	Petiole pigment	Root shape	Skin colour	Flesh colour	Extracts of farmers' characterizations
klarin	7	1	2	5332	32	1	8	830	100	grows easily; dry, sweet roots; tops for veg, forage; saleable.
5-fingers	7	3	5	6755	72	3	3	634	442	quick maturing; leaves for vegetables; sweet, watery roots; saleable ('*comote cue*').
amerikano	7	7	8	3131	95	9	8	126	543	profuse vines; few big roots; sweet, powdery.
igorot puti	9	3	3	5332	62	3	8	230	200	profuse foliage, good for weed control, forage; roots sweet, dry.
igorot pula	7	3	5	5334	92	8	3	230	100	dry, sweet; saleable in market; leaves for vegetables.
tapol	5	7	8	3151	95	9	2	930	196	easy to maintain; unsweet, watery roots; vegetable; good market for this colour; small area.
kamada	5	1	2	5332	23	3	3	520	100	quick maturing; easily harvested; sweet, watery; suitable for forage.

Notes on morphological descriptor codes

1. Plant type: 1 erect 9 extremely spreading
2. Vine colour: 1 green 9 totally dark purple
3. Leaf shape: outline 1 rounded 7 almost divided; type of lobes 1 none 9 very deep; lobe number; shape of central lobe 0 absent 9 linear
4. Immature and mature leaf colour: 1 yellow-green 9 totally purple
5. Root shape: 1 round 9 long irregular or curved
6. Skin colour: main 1 white 9 dark purple; intensity 1 pale 3 dark; secondary 0 absent 9 dark purple
7. Main flesh colour: 1 white 9 strongly pigmented; second. colour 0 absent 9 dark purple; distr. of colour 0 absent 9 covering all

of the garden intervened at this point, seeing the abandoned garden as an unusually rich source of novel planting material for her own fields. Following earlier practice, she made an initial selection of ten sweet potato cultivars from the original collection which 'she liked' in an as-yet unspecified way: *sugahak, amerikano, turay, si-uron, kapayas, kalibre, igorot puti, igorot pula, 5–fingers* and *sil-ipon*. They were planted in a small plot near her house to produce planting material for a planned larger site. Seven varieties of those rescued were finally transplanted to that site: *si-uron, igorot pula, igorot puti, 5–fingers, amerikano, turay* and *sil-ipon* (Table 6.4). Why these seven? Because, she explained, they had produced enough planting materials. The others, with little plant growth, were deselected by default.

This woman expressed interest in obtaining new varieties because of plans to start pig-raising, which would explain the interest in the highly spreading variety *igorot puti*, but she also cited several important agronomic traits that she evaluated, such as earlyness, ability to survive under stress, high root yield and a 'stomach filling' capacity that can substitute for rice or corn. Since no one cultivar will satisfy all these requirements, her aim in maintaining the new collection was 'conservation for evaluation'. Having already selected out less robust cultivars during the production of planting materials, she wanted to further evaluate the seven accessions from the original site plus some additional cultivars of her own. She compared cultivars long known and grown in the locality, such as *turay* and *si-uron* with lost cultivars like *sil-ipon*, remembered from the 1950s or recently acquired varieties, such as *igorot pula, igorot puti* and *amerikano,* which were brought from the Maambong conservation site, about one-and-a-half hours distant.

The selected set closely resembles the cultivars that emerged as dominant in the Maambong site. Some are the same genotypes *(igorot puti, ingorot pula and amerikano*), but others are what we can call 'isomorphs': they are genetically different but have very similar characteristics and functions (for example, *gireng* and *five-fingers*; *tapol* and *turay*). This suggests that there is a regionally common clustering of varietal needs which are met by sets of different genotypes with very similar overall genetic characteristics. Thus, unlike in some other crops, such as potato, where 'regional varieties' emerge to satisfy particular needs (cf. Brush *et al.*, 1981), regional needs for particular types of sweet potato, and perhaps for taro too, are met in this part of the Philippines by a series of local isomorphs. The implications of this for local, farmer-based crop genetic conservation are discussed in the conclusions.

What can we say about the women's management of genetic diversity in relation to the 'residual cultivars'? The deselection of cultivars that failed to provide enough planting material, referred to above, already gives an idea of the stringent test of survivability applied to this crop. But the actions of the Dalwangan farmer also clearly indicated a curiosity to test new material. A closer look at the cultivar maintenance activities of one of the women involved in the Maambong conservation can provide further illumination of the fate of the residual cultivars.

Mrs. Fely delos Santos explained that every time she goes to a particular area, she would observe their sweet potato varieties and ask for some cuttings if she feels that they are different from those she has in her collection. She ignores other factors that are often taken into consideration in planting sweet potato, like ability to produce plenty of roots, deliciousness, acceptability in the market, and the like, as long as it is different from what she already has. This woman has

Table 6.4. Characteristics of seven cultivars selected for maintenance by Mrs. Zagado, Dalwangan

Cultivars	*Plant type*	*Vine colour*	*Leaf vein colour*	*Leaf shape*	*Imm/ mature leaf col*	*Petiole pigment*	*Root shape*	*Skin colour*	*Flesh colour*	*Farmers' characterizations*
*si-uron**					'green/ light g'		'elliptic'	'red'	'white'	clustered, profuse roots, sweet, powdery.
sil-ipon	7	3	2	3111	62	5	1	230	100	few roots, bland, watery.
amerikano	7	7	8	3131	95	9	8	126	543	profuse vines; few big roots; sweet, powdery.
turay	7	6	8	3111	95	9	9	930	199	leaves for vegetables.
igorot puti	9	3	3	5332	62	3	8	230	200	profuse foliage, good for weed control, forage; roots sweet, dry.
igorot pula	7	3	5	5334	92	8	3	230	100	dry, sweet; saleable in market; leaves for vegetables.
gireng/5-fingers	5	1	7	6955	62	4	9	230	100	

* No technical characterization data available because it was lost from *ex situ* genebank

consistently maintained the greatest diversity of the different rootcrops, ranging, for sweet potato, from 8 to 12 varieties over the five plantings. At the same time, many of these 'different' cultivars are planted in only a few hills and disappear from her plot after one or two replantings (Table 6.5). Both from her testimony and her actions, there is no doubt that Mrs. Santos is collecting diversity. But it also seems that new cultivars need to satisfy very clearly particular needs better than an existing preferred cultivar for it to be selected. This occurred in the case of selections made by Mrs. Zagada in Dalwangan. In Maambong it has not yet occurred. It will be helpful to examine in more detail the attitudes towards evaluation and selection in Maambong.

Evaluation and use of diversity

We need to recognize that in most cases, local crop diversity is maintained for particular uses. Though we have stressed in our own approach the development of a public culture of conservation as something that can enhance the long-term sustainability of systems, we have also recognized that local conservation will be genetically dynamic rather than static, with novel cultivars (naturally produced or introduced) being evaluated for their 'fit' with local systems, leading to additions of positively evaluated cultivars, and inevitably the disappearance of those negatively evaluated or considered residuals. This is a good reason in favour of local conservation gardens – their capacity to generate new land-races through natural processes and the evaluative and selection procedures of local experts.

Although, as we have seen, some members of the conservation group in Maambong actively sought out novel cultivars early on, most of the cultivars

Table 6.5. Variety maintenance by Mrs. Fely delos Santos, Maambong conservation garden.

Cultivars	*1st Planting*	*2nd Planting*	*3rd Planting*	*4th Planting*	*5th Planting*
klarin	**x**	**x**	**x**	**x**	**x**
5-fingers	**x**	**x**	**x**	**x**	**x**
amerikano	**x**	**x**	**x**		
igorot puti	**x**	**x**		**x**	**x**
igorot pula	**x**	**x**	**x**		**x**
kamada	**x**			**x**	
topol					**x**
kinampay	x	x	x	x	
bilaka	x	x	x		
valencia		x	x		
magtuko		x			
maranding		x			
kapitlok		x	x		
senorita		x			
kaligatos			x		x
kabato				x	
kawakwak				x	x
PNGL				x	x
P16				x	x

Note: dominant cultivars in bold

initially planted by the individual members were already known and therefore not subject to special evaluation. Later, novel cultivars were introduced into the conservation garden as a result of cross-visits to other sites and to a regional genebank. Women selected particular cultivars because leaf shape or vine pigmentation were similar to a known variety, or the variety appeared 'robust' or healthy and attracted attention. Once the cuttings were brought home and planted in the local garden, few of the women seemed to take great interest in the agronomic performance, though morphological and growth characteristics were observed. For many of the group, there was no real excitement at the harvest either, more a matter-of-fact observation of the root production. Roots of the new cultivars would be tried out, and the taste characteristics noted and it was easy to elicit these characteristics from the women. However, none of the newly introduced cultivars appeared to achieve the status of dominant cultivar during the five seasons and, as we have seen, most dropped out after one or two plantings. Thus, the six to nine cultivars each woman is maintaining by the fifth planting consist of between five and seven dominant varieties with additional 'different' varieties either being evaluated or planted because planting material happened to be available. If only a few cuttings are obtained, it is likely that the cultivar will disappear after one or two replantings.

To illustrate this rather casual attitude towards the evaluation of new sweet potato varieties, we can consider the request made by group members to the research team to help them obtain 'carrot-like' varieties of sweet potato, which did not exist either locally or regionally. In response to the request, researchers introduced four exotic cultivars into the Maambong site for evaluation in the fourth planting: P16, a selected local cultivar from the Mount Pinatubo area of northern Philippines; NPSP, a local selected cultivar for the northern mountain area of the Philippines; UPLSP, a Philippine Seed Board Variety, and PNGL, a variety from Papua New Guinea. P16 was given to all women, but the rest were distributed according to availability of space, since other cultivars had already been planted.

Out of the 13 farmers with plots in the fourth planting, only four still had P16 during the evaluation at the time of harvest. Two still had UPLSP and one had PNGL. Many of the UPLSP and all NPSP cuttings died. Both were found to be very susceptible to scab. The carrot-like P16 survived better and was valued for early maturity. The four members who were able to harvest it replanted in the fifth planting and some cuttings were given to other members. Why were nine members unable to harvest P16? To a large extent it was due to the small number of cuttings planted and their vulnerability to weeds and drought. No special attention appeared to be given these cultivars. They were allowed to flourish or perish. Perhaps this explains the approach to cultivar evaluation. Sweetpotato is considered an easy-to-manage crop and in the farmer evaluations of varieties in Tables 6.3 and 6.4, ease of management is an important evaluation criterion. Survivability is perhaps the first and most crucial characteristic that cultivars have to show and the women's hands-off approach is the best form of evaluation.

Conclusions

Maintenance practices and outcomes

This chapter has not dwelt in any detail on the action research component of the project, which explored ways to support local cultivar management and therefore it is not, in general, possible to draw conclusions about that aspect of the work. Nevertheless, on the basis of evidence presented here, the project seems to have contributed to an overall increase in the diversity managed by farmers in at least two sites. This can reasonably be interpreted as due to an increase in sensitivity to conservation of rootcrop diversity and perhaps also, a recovery of diversity management strategies which were disappearing.

The study also confirmed that there is considerable dynamism in the local gardens, with many additions and losses of accessions. On the basis of the *ex situ* evaluations of the material held in the different sites, the diversity seems quite high, or at least, there is quite a wide range of morphological characteristics in the accessions which are consistently maintained. On the other hand, a large number of cultivars seem to be very close genetically and it may be that genetically similar accessions are allowed to disappear from individual collections with little concern. More work on the *ex situ* site is needed to confirm this observation.

The cultivars maintained in the conservation gardens are still largely local and are far short of the range of cultivars found in the province of Bukidnon. Relatively little material has been introduced and maintained by farmers following their visits to other areas. It was very clear that whereas many women were interested in collecting new cultivars, this did not necessarily mean they planned to conserve them. Though most women readily characterized new accessions in morphological and agronomic terms, it seems that the first basis for evaluation is survivability. Only then do other evaluation criteria come into play. What we have seen is that local people tend to conserve a core set of 'dominant cultivars' which score highly on several specific criteria of evaluation for satisfying particular needs. The set of these cultivars appears to capture a large part of the genetic diversity of the province, if not the genotypes. We have also found that although the set of dominant cultivars differs from site to site, they seem to possess similar characteristics. They could be described as 'isomorphs' of each other.

At any one time, several of the cultivars being grown by farmers – almost always represented by very few plants – are really residual, readily being allowed to drop out of cultivation after one or two seasons. The vulnerability to loss of these less common sweet potato cultivars planted in only a few hills echoes a trend identified in the Andes of Latin America for certain native potato varieties (Brush, 1992). Although many commercializing farmers in the Andes continue to plant a wide range of local varieties, they do so on smaller plots in more marginal areas than previously, reserving an increasing area of the better land for the high-input cultivation of modern varieties. This makes it more likely that native varieties disappear. However, we should be careful not to take the equation of low plant density with likelihood of erosion too strongly. Within the set of 'dominant cultivars' there is also variation in plant density. Some are planted in quite limited areas to satisfy specific, but limited, needs. This suggests that area alone may not be enough to determine whether a cultivar will disappear, if it has a clear use within the local livelihood system.

These findings are still provisional and there is a need to confirm, either through further morphological evaluation or through molecular techniques, the genetic diversity of the dominant cultivars in relation to the genetic diversity of the region as a whole. If these findings are confirmed, it suggests that a number of communal gardens could adequately manage the broad range of genotypes present in the region with the possibility that other genotypes could be conserved as botanical seed in an *ex situ* facility.

Livelihood and sustainability

The findings clearly relate local management of diverse cultivars with the existence of a diversity of uses for the crop. To ensure the sustainability of local conservation efforts, they must be combined with livelihood opportunities on the one hand, and on the other, local maintenance of crop diversity needs to be integrated into a national strategy for the conservation and use of crop germplasm. Both the livelihood component and the institutional integration would contribute to the legitimization of local efforts; in other words, to developing the public culture of conservation which is also a necessary, but not sufficient, component of successful local management of diversity. Integration within the national system could provide the kind of long-term support which this project has provided on a short-term basis through the on-site visits by the research team.

Although not discussed in this abbreviated account of the project, we have identified a number of opportunities for integrating livelihood elements with local conservation. By institutionally linking conservation with home gardening, a food security focus is introduced, which emphasizes diversification of crops and cultivars within the site to satisfy a diversity of food and nutrition needs. A complementary approach, discussed but not yet implemented in any of the sites, is the diversification of market niches for different cultivars of a crop through developing different uses, beyond the fresh root market, the forage market and the use of tops for vegetables already described. Such an approach can also be combined with other initiatives on the genetic resources side. Instead of depending on available local germplasm as a potential source of raw material for new uses, participatory plant breeding can offer the opportunity to cross dominant local cultivars with advanced material from elsewhere, so as to expand the amount of promising material available for evaluation (Joshi *et al.*, 1996; Sthapit *et al.*, 1996; Whitcombe *et al.*, 1996). Such an approach, however, would not be appropriate in all locations. The potential for the crop to make an important contribution to livelihood and the existence of high potential alternative uses will be determining factors.

Acknowledgements

The authors wish to acknowledge the contribution to this work of Dr. Virginia Nazarea, who was leader of a project which rigorously documented the indigenous knowledge of rootcrop diversity in Bukidnon, southern Philippines and constituted the first phase of the present activity.

Ms. Lilibeth Zagado was involved both in that earlier documentation work and as research assistant on the conservation activities reported here. She played a very important role in this study.

Technical advice and support received from Dr. Jose Bacusmo of the Philippines Root Crop Research and Training Center is gratefully acknowledged, as are the incisive comments of Professor Gelia Castillo on earlier versions of this chapter.

Finally and most importantly, special tribute is paid to the participation – often enthusiastic, sometimes long-suffering – of the farming and gardening groups in the four sites in Bukidnon, many of whom continue to be involved in rootcrop conservation work. This chapter is dedicated to them.

7. Farmer experimentation in a Venezuelan Andean group

CONSUELO QUIROZ

Abstract

THIS IS A case-study of a group of resource-scarce farmers in an Andean region of Venezuela. Topics studied include: incentives to experiment; characteristics of main experiments; agricultural knowledge. The study shows that experimentation is one of the major activities that farmers use for adapting to their surroundings and to changing conditions; trials are not random, and are geared to what farmers can manage; trials are closely monitored and evaluated. The experiments include spiritual aspects, and it is important to understand the farmers' world view.

Introduction

An extensive literature exists about agricultural innovation and diffusion around the world. However, very little attention has been given to local farm-level innovation and still less to farmers as active experimenters or innovators in developing countries (Haverkort *et al.*, 1988; Johnson, 1972; McCorkle, 1994).

Small-scale farmers are often said to have conservative attitudes towards change and innovations and to be passive, 'programmed by their cultural learning to respond to finite and discrete sets of environmental conditions' (Johnson, 1972: 153). In the last 15 years there has been an accumulating body of evidence that proves that these assumptions are wrong. Farmers are not just 'clients' of 'change agents', they have been active as self-directed learners in developing technologies for the production, processing and storage of food since the earliest stages of agriculture (Berkes and Folke, 1994; Haverkort *et al.*, 1988; Hyndman, 1992; McCorkle, 1994; Rhoades and Bebbington, 1988; Warren, 1991).

The majority of small-scale farmers in developing countries live under conditions of high variability, uncertainty and complex interactions (Altieri, 1987). For them, the indigenous innovations generated mainly through experimentation represent one of the key elements for their survival (Reijntjes and Hiemstra, 1989). Small-scale farming could well be called, as Reijntjes and Hiemstra say, 'the world's largest research laboratory' (1989: 4).

Unfortunately, an important number of indigenous technologies and knowledge systems relevant to sustainability are being lost at a very fast speed (Cunningham, 1991; Haverkort and Millar, 1994; Mathias, 1994; UNEP, 1994; Warren, 1992). Many traditional practices have been eroded and others are not performed with the same effectiveness due to a complexity of factors that include European colonization, the neo-colonization process and the Green Revolution (Gliessman *et al.*, 1981; Shiva, 1993). There are many development institutions that are seriously engaged in replacing indigenous technologies with ones that are considered more 'modern' and more 'productive'. In the case of the Andes and most of the rural areas in developing countries, this has proven to be a wrong approach, because for more than 40 years these attempts have only caused mass poverty, hunger, more food imports and environment deterioration (Rengifo-Vasquez, 1989; Shiva and Dankelman, 1992).

The purpose of this chapter is to contribute to the understanding of the farmer experimentation process. This will be done through the following objectives: (1) presentation of an overview of recent literature on the topic, (2) presentation of a case-study of farmer experimentation from an Andean region in Venezuela; (3) drawing out the implications and observations concerning the role of farmer experimentation for the extension/development practitioner's work.

Farmer experimentation

Although farmer experimentation has been almost totally ignored in the development literature, there is strong empirical evidence that shows that farmer experimentation is a very common phenomenon in small-scale agriculture in developing countries (Johnson, 1972; McCorkle, 1994; Reijntjes and Hiemstra, 1989). For example, in a survey made by Rhoades and Bebbington (1988) it was found that 90 per cent of all settler farmers of the upper Chanchamayo in Peru were avid experimenters. 'Perhaps some were more active than others', the authors pointed out, 'but virtually all conducted "*pruebas*" or trials, particularly with potatoes' (1988: 3). This does not mean that all farmers are innovative and are able to cope with changing circumstances. There are other constraints apart from the technological ones which are frequently more limiting (such as the social, political, cultural and economic context) (Reijntjes and Hiemstra, 1989).

The concept of experimentation

Although scientific researchers use the term 'experiment', farmers seem to prefer the term 'trial' (or its equivalent in their own language). For example, in Spanish they use the word *prueba* (Rengifo-Vasquez, 1989) and in Mali they use the word *Shifleli* (Stolzenback, 1993).

This process is not one of mechanistic repetition, but rather one of re-creation. It is a permanent type of activity which is directly associated with the social and productive processes (Rengifo-Vasquez, 1989).

Major steps in trials

Anthropologist Constant McCorkle, in her study of farmer innovation and diffusion in Niger, pointed out that the major steps that farmers take in designing, implementing, and evaluating their trials are the following (1994: 33–4).

1. Gathering background information from farmer-colleagues and other knowledgeable persons and from direct observation of others' experiences with field-trial design and the variety to be tested. In addition to that, they do an oral 'literature review' prior to initiating an experiment.
2. Selecting the field-trial sites according to consciously established criteria, e.g., depahic and hyrologic characteristics.
3. Controlling for major variables. In the cases recorded in her study, these most often focused on use versus non-use of manure or chemical fertilizer, size of experimental plots, and planting densities.
4. Running the trials for more than one year to allow for differing performance results due to inter-annual climatic and other variations.
5. Monitoring and evaluating the results of trials according to the features of interest are standard features of the trial process.

Characteristics of farmer experimentation
Several common characteristics of small-scale farmers' experimentation can be identified (Bentley 1990; Haverkort and Millar, 1992; Rengifo-Vasquez, 1989; Rhoades and Bebbington, 1988; Scoones and Thompson, 1994; Stolzenback, 1993).

Holistic, on-going process. Farmers' performance occurs embedded in a particular agro-ecological and socio-cultural context that exists beyond the farm-gate and that is usually beyond the farmer's control. This gives rise to a variety of changing conditions to which the farmers must make a series of adjustments. Experimentation represents one of the major on-going activities the farmers use for adapting to these changing conditions. It involves all the elements of the peasant life system. As Rengifo-Vasquez argues, 'the farmer is interested not only in the behaviour of a crop, but also in its relationship and balance with the other farm's elements (crop, animals, society)' (1989: 10).

Choice of 'the right moment' for trial. Trials are not performed as random activities. There are circumstances or 'right moments' where, according to the farmers, the conditions are 'right'. This type of behaviour is related to the comprehensive knowledge they have about the interrelationships that exist among the different elements of the vital cycles of nature. For example, in the case of crops, these elements would include the soils, water, climate and seeds.

Considerations of religious and other beliefs. Farmers' experiments usually include spiritual aspects. These aspects are often unperceived or even ridiculed by outsiders, but they are very much a reality for the people who believe in them. This is what Haverkort and Millar call 'cosmovision'. Cosmovision, they argue,' 'assumes interrelationships between spirituality, nature and mankind. It describes the roles of the superpowers and the way natural processes take place' (1992: 26).

This cosmovision is reflected in behaviours that include, for example, 'no planting before certain religious festivals have taken place', and carrying out some of the agricultural activities according to the phases of the moon.

Size and number of trials. The size of the trial is usually a size the farmer can handle ('manageable size'). It consists, often of only a few plants or animals. If the first results from the trial are positive, then the crop is planted in a bigger area (or the treatment is applied to a larger number of animals), unless this is restricted by factors such as the market and seed supply. But if the trial doesn't work, the farmer usually continues trying, because he or she wants to know the causes of the failure. A farmer explained to Rengifo-Vasquez that 'every crop looks for it own type of soil, climate and treatment' (1989: 12).

Location. Trials are generally located in places chosen on purpose. The trial usually is not placed alone but with other plants which are already adapted.

Care and love. The new plant or animal is handled in an individualized way. In other words, it is treated with all the necessary attention, as if it were a 'new member' who is coming to the family. The Aymara ethnic group from the southern Andes, for example, has a name for this new member, *yockcha-nuera*.

This term refers to the condition, as a 'relative', that this new member has for the farmer (Rengifo-Vasquez, 1989: 12).

Experimentation is not equivalent to risk. Experimentation and risk are separate matters. Small-scale farmer experimentation is a good example of a situation that shows it is possible to experiment at a 'low cost and low risk'. It is correct that traditional agriculturists are 'cautious' in the face of innovations; but it is not true to say that they refuse to try new things.

Important role played by older persons. The innovation process, according to the 'modernization' literature, is dominated by young persons. In the case of farmers' trials it is usually those with the most experience, generally the older ones, who can perform the task better and who are recognized as the more knowledgeable about that particular crop or livestock.

Gender and experimentation. Women in many parts of the world have traditionally played a very important role, not only in the conservation and enhancement of genetic resources (biodiversity) but also in agricultural production in general (Quiroz, 1994). This knowledge, and their active role in experimentation, has allowed them to develop and utilize what some people call a 'science for survival' (Rocheleau, 1991).

There is a growing number of empirical studies that show that there are gender-based differences in local knowledge (Badri and Badri, 1994; Norem *et al.*, 1989; Rocheleau, 1991; Shiva and Dankelman, 1992). Some authors even talk about women's knowledge as a 'distinctive' type of knowledge (Jiggins, 1994). Therefore, it would be incorrect to assume always that the male farmers' experimentation process is similar to the women's.

Types of farmers' experiments

In a study made by Rhoades and Bebbington in two Peruvian potato production areas, farmers' experiments were characterized as: (1) curiosity experiments; (2) problem-solving experiments; and (3) adaptation experiments (1988: 10–14).

The curiosity experiment refers to those experiments set up by farmers to test an idea that comes into their mind. These experiments may or may not have an immediate practical end. Farmers frequently develop ideas for experimentation that seem strange to scientists. Problem-solving experiments refer to those experiments set up by farmers to seek practical solutions to new and old problems. Adaptation experiments are conducted by farmers after they acquire a new technology, or after they have observed a new technology demonstrated elsewhere. Such experiments can occur in three contexts: (1) when farmers are testing an unknown component of a technology within a known physical environment; (2) when farmers are testing a known technology within an unknown environment; and (3) when testing an unknown technology within an unknown environment. The questions that farmers expect their trials to answer are: (1) does the technology work?; (2) how can it be fitted into the existing production-utilization system?; and (3) is it profitable?

Evaluation criteria in farmer experiments

Farmers' quantitative analyses are based on their observations. The strong point of farmers' perception, and therefore their evaluation, is their frequent observation and monitoring during the entire trial process. They carefully check the trial through close observation (and sometimes written records) of the performance of the new variety or animal introduced. Retrospectively, farmers can determine factors that could have influenced the yield. The parameters often used as evaluation criteria vary according to the variables being evaluated. For example, in the case of crops, these parameters include, vigour and colour of the stem, colour of the leaves, height of the plant, size of seeds, moment of germination, ripening, resistance to drought and high yields. (McCorkle, 1994; Stolzenback, 1993). The running and close monitoring of a trial usually gives the farmer enough information to reject the technique or try it out the next season on a larger field, possibly under slightly different circumstances (Stolzenback, 1993).

Case-study

The empirical information for this case-study is based on fieldwork done in the sector Loma Gorda, Trujillo, Estado Trujillo, Venezuela during the end of 1993 and beginning of 1994. This study presents some preliminary results coming from an on-going project regarding the documentation of indigenous agricultural knowledge carried out at VERSIK (Venezuelan Resource Secretariat for Indigenous Knowledge). The information was collected by using ethnographic research methods, mainly semi-structured in-depth interviews and participant observation. Fifteen interviews and observations were made with nine producers.

Lorna Gorda has a population of 90 inhabitants living in approximately 15 households. It is a rural agricultural area located 90 minutes by car from Trujillo, the capital of Trujillo State.

The major crops in the area have varied across time. Until the 1950s sugar cane was the most popular crop in the area. Some of the farmers even had in their homes the necessary equipment to process the cane and prepare 'brown-sugar blos'. At the end of the 1950s, coffee production was introduced and most of the farmers switched to this crop because it offered more advantages from the economic point of view. This crop is still grown in the area, but in very small quantities due mainly to relatively recent marketing problems. There are other crops that have been traditionally cultivated mainly for subsistence purposes, including maize, black beans, cassava, plantain, bananas and oranges. In the late 1970s a farmer from Colombia came to the area and started to grow some horticultural crops. He began with potatoes, green peppers, coriander, lettuce and green beans. For several years he was the only one in the area cultivating that kind of crop, but as the years passed Lorna Gorda's farmers observed what he was doing successfully and then some of them in the early 1980s began to grow these crops. Today this is the major source of income for farmers in this area.

Characteristics of farmers and farms

All the people interviewed began their careers as farmers when they were children. They recognized their fathers as being their major instructor in agriculture. For example,

> I began as a farmer when I was 7 years old. My father taught me the things I know. He used to have meetings with his workers and I was one of them. I began growing mainly maize, cassava, coffee, black beans, guand and bananas. (farmer 3, int. 1).

Their average age was 42 years, the youngest being 30 and the oldest being 60. Only two did not know how to read and write, the rest having had some years of primary school (some completed it). Two of them finished secondary school.

Only one of the farmers did not own his land; he rented it. The average size of each lot was 6.2 ha. Seven of the farmers had land which ranged in size between 3 and 6 ha, and only two had more than 10 ha (12 to 13 each). However, these two farmers were not able to use all their land because their plots had large areas of unusable land due to very steep slopes.

The labour force was supplied mainly from their own family. On some occasions, especially during harvest time, they either hired additional workers or made use of a co-operative type of work system called 'returned-hand' (*manouelta*).

Farmer experimentation. The term utilized by these farmers is *prueba*, which is the equivalent to 'trial' in Spanish. They call trial that activity where they are introducing something totally or partially new in their farms, such as a crop, a variety of a crop, or a different management system. The main role played by the trial is to 'try to avoid losses (in terms of time, money and/or labour).' For example,

> You need to try first because otherwise you may get 'broken' [Ud. se '*envaina*']. You have to try first, you don't know yet whether the crop will do well or not. If you don't try and the crop doesn't go right you may have losses regarding fertilizer, labour and you even lose the 'sweat' you put on it. You may lose a lot of money and when you don't have enough money you cannot afford to lose it.

Trials as an 'on-going' and diverse kind of process. They are constantly actively engaged in conducting a wide diversity of trials. They try, for example, different kind of crops, different varieties, different amounts of fertilizers, different planting distances, different planting times and different soils. For example,

> ...I have tried with potatoes and green beans. They both grow very well over here. I am going to try with cucumber. So, I asked my companion to bring me a can of seeds, to see whether they grow well over here. I will plant just a few of those seeds down there, to see how they do. (Farmer 3, int. 2).

> ...I have tried different kinds of bananas, for example, manzanito, topocho, dorao, titiaro, manchoso. Those kinds are not done good in this area, but coco valenciano is very good for this area... (Farmer 2, int. 1).

> ...Not every land will produce the same. Sometimes you grow some seeds in a fallow land and you get a good harvest, but next season, let's say there is a strong summer, lack of water or too much rain, then it is not the same. These are the experiments that we farmers have to do. You win some and lose others on the same land. For example, I have tried cassava in that piece of land, and it did not work, and I have tried in this one with maize, and it did not work either, but this is a very good land for black beans. (Farmer 1, int. 1).

Type of experiments

Examples of the three types of experiments identified by Rhoades and Bebbington (1988) can be found in the area.

Curiosity experiments. These occur when the farmers are given any new kind of seeds by neighbours or friends; they usually try them out on their land, even if they do not know what behaviour to expect from them or where the seeds are coming from. They try them 'just to see what happens.' For example,

> ...I was given around 1/2 kg of maize seeds called 'Peruvian maize'. I tried them. They grew very big. I watered them frequently. That was really a good experiment. (farmer 4, int. 1).

Problem-solving also occurs. One of the problems most frequently mentioned was the 'lack of money'. Therefore, whenever farmers know about a crop which has a good price in the market, that is an incentive for them to try it.

> ...when the people say that a crop has a good price in the market, that is one of the reasons why we try it. We are always trying to look for ways to make money, and if we have never tried it, then were go quickly and try! (farmer 8, int. 1).

> ...when I learned that flowers have a good price in the market I tried them in my backyard garden, and they grew very well. Then I tried 50 plants and they also grew well, so I decided to plant 400 plants to see what will happen. (farmer 2, int. 3).

The second most important reason mentioned to experiment was the need for a particular crop either as food for self-consumption (such as subsistence crops like maize), or as a medicinal plant (e.g., aloe). Experimenting with and growing these kinds of crops mean that they would have those crops available without needing to buy them in the market. Examples include:

> ...ginger, this a very good medicine when you have cough, that is why I am trying it. I brought a plant from Trujillo because if I can grow it I won't need to buy it anymore. (farmer 7, int. 1).

> ...if I like to eat the product, for example a kind of beans, then I will experiment with it several times. (farmer 2, int. 3).

There were other kinds of problem-solving mentioned. For example, a farmer decided to try pines to protect some steep slopes in his land he knew were very eroded due to many years of slash and burn management.

> I am really 'in love' with those pine trees. Some people think I am growing them to make money, but that is not true. Who has ever seen an old man [he is 60 years old] like me planting pines? Those trees do not produce any fruits. That is not my case. I first asked for 1000 plants of pines to try them. They were great! Then I was motivated to plant more of them since I have some pieces of unusable land, very steep slopes, where I cannot grow anything. That land has been burnt many time. So I decided to try pines because they protect my land. (farmer 5, int. 2).

Adaptation experiments. These are undertaken, for example, when a farmer observes that another farmer is growing a crop which is doing well on that farm, and is motivated to try to adapt it to his or her farm conditions. These trials provide the farmer with confidence regarding the possible results from the experiment by testing an unknown technology in a known physical environment.

> I tried potatoes because another producer who lives in the upper area had tried them and had good results. So I tried and has as good results as he had told me I would have... (farmer 3, int. 2).

'Double experimentation'. This is a variation of any of the kinds of experiments already mentioned, and refers to the cases where the farmers are consciously trying two variables at the same time such as different fertilizers with different planting distance. An example was a farmer who talked about the different kinds of fertilizers he was testing for growing potatoes,

> Now I am going to try putting a double amount of fertilizer, and I will plant each plant at a larger distance one from the other. Do you know why? To have bigger potatoes. To grow potatoes is something very difficult; you need to know about it, otherwise you lose the harvest... (farmer 3, int. 2).

Number and size of trial repetitions

If the first trial has the expected positive results, then the farmer may decide to increase the number of plants, and consequently the area of land. If the first trial did not work, then the farmer may decide to reject that technology component (e.g., a variety of a crop) or to try again two or three times, making some changes. The reasons to try again are generally to find out what caused the failure of the first trial. The number of repetitions of the trial will depend on factors such as the availability of seeds and the interest the farmer has in that crop,

> ...I do not like when I try something and it does not grow well, but you have to try again, because you need to know that is going on, was the failure due to fertilizer?, to the amount of water?, to the climate?, etc. I always do 2–3 trials (farmer 3, int. 3).

> ...if the results from the trial are not good I always repeat it. I never experiment only one time, unless I am unable to find more seeds. (farmer 2, int. 3).

Location of trials

Trials are neither separated nor marked in the agricultural plot, and occupy only one small portion of the whole. The trial is usually planted where that same crop would be grown if the trial were successful. The rest of the plot is planted with a common crop.

> ...of course I would do the trial in the same fallow land that I would use if the trial works out well, otherwise it would not have sense. Another piece of land would not be the same; all the plots are not the same. Some have some 'foods' [needed for the plant] and others have different ones. In the rest of the land you plant another crop you know grows fine because you cannot afford to have a piece of empty land. So the next season if you like what you were trying out, then you plant all the plot with it, and if it has good price in the market, then you make a lot of money. (farmer 3, int. 3).

Trials are located in any area (corner, centre, etc.) of the chosen plot. What seems to be very important is that it is located in the same plot that would be used if the trial outcomes are positive. For example,

> ...you choose a small pieces of land for the trial form anywhere. Let me explain it. For example, it can be in the middle, on the border, any place. What is very

> important is that it is the same land where you would plant it if the trial was OK. (farmer 7, int. 1).

Farmers usually refer to the size of the trial as something 'small'. One way to get an idea about the size is to know the number of seeds they would regularly use in the trials. It was found that it will depend on the type of crop they are trying. Seeds are classified into two groups: big and small. 'Big' are crops such as avocado, berries, citrus and plantain. With 'big' ones they utilize a very small number of plants, for example two or three. Sometimes they consider trying where only one is enough (e.g., only one 'passion-fruit' tree). With 'small' seeds, for example, maize, beans, peppers, tomatoes and lettuce, they utilize larger quantities. For maize, beans or green beans, they utilize 0.5–1 kg of seeds. If it is peppers, lettuce, tomatoes, cabbage, they use a can of seeds (approximately 200 plants). For example,

> ...the amount of seeds you use will depend on the crop you are trying. For example if you are trying maize, whose seeds are very small, then you will use 0.5–1 kg for your trial. You cannot try with 1–4 seeds as you do with other crops whose seeds are larger, such as oranges and avocados. In this case, a small amount of plants will be enough to know whether they will 'do well or not'. (farmer 5, int. 2).

Religion and other beliefs

It is believed that to work during the days considered 'Holy days' is not good, because if you do it then 'God won't help you'. For example, they usually do not work during Holy Week (Easter), Cross day (May 7) and Corpus Christ (June 17). They believe also they should not plant anything during the month of November because this is the month of the 'souls' (dead people). For example,

> ...you should not plant anything during the 'souls month' [November] because if you do, nothing will grow. You have to pay respect to God during those days... (farmer 8, int. 1).

> people who work during the 'Holy week' are those who are ignorant, who does not believe God exists. That is why everything they do goes wrong. (farmer 9, int. 1).

The influence of the moon is recognized by the majority of farmers in the area. They know that the lunar phases influence the plant's life cycle (e.g. planting, growing, production, pruning, harvesting) of various crops such as cassava, plantain, beans, maize and coffee. They are aware of the existence of this influence not only because their ancestors have told them so, but also because they themselves have made trials about it. For example,

> ...if a farmer wants to plant cassava, it has to be done during the last week when the moon is growing and the first week when the moon is disappearing. Farmers know those things. If you plant when there is no moon, nothing will grow well, whether it is cassava, maize, beans, anything. These are the kinds of things we farmers know because we have experimented with them. (farmer 2, int. 2).

> I have planted bananas and maize during the 'moon change' phase, not in big amounts, but in cases where there is a small piece of land without anything on it. I have planted them in place, and they haven't grown well. You learn from

> the experience and realize that you have 'violated the rules'. Every time you get more convinced that it is necessary to follow what the older persons tell you to do. I have experimented with everything and everywhere. (farmer 4, int. 2).

> . . . the moon influences everything you do with the plants. I have tried these things. For example, when you are pruning a tree, let's say an orange or an avocado tree, if you want that tree to produce big and healthy fruits, you have to do it when there is a full moon. (farmer 9, int. 1).

One of the farmers believes that moon phases influence only the cycles of some crops:

> ...in relation to maize, horticultural crops and black beans, you don't need to take into account the moon phases. But with other crops like cassava, plantain, bananas, you do have to take it into account, otherwise you don't get good results. It is necessary to be careful. For instance, it is necessary to know that you have to cut the cassava seeds three days before there is a full moon. (farmer 5 int. 1).

The majority of the farmers interviewed also believe in the influence of what they call the farmer's 'good or bad hand'. This 'hand's influence' may affect several farm operations, e.g. cutting the asexual seeds, planting and animal castration. For example, it is believed that a person who is considered to have a 'bad hand' should not take part in the preparation of seeds because if he or she does this, the seeds won't yield good results. For example,

> ...I really can talk about 'good or bad hand'. There are some people who have 'good hand' to grow something and 'bad hand' to grow other things. I noticed that when a man came here and cut my cassava seeds, he had 'bad hand' and then the plants didn't grow well. I thought at the beginning that it happened because the place where I planted them. But I tried again in the same place, this time cutting the seeds myself, then I realized that what happened was not because of the seed, or the location, it was because the man's 'bad hand'. (farmer 1, int. 1).

One of the farmers interviewed doesn't believe in the influence of the person's hand. He thinks the important thing is to know how to perform the work well. He said,

> hand beliefs are not true. That doesn't exist; what really exists is the 'bad memory' and the 'bad understanding'. If you give me an advice, because you have had experience on that and I do not accept it and I say that you do not know anything because the one who knows is me, then whatever I do is not going to go all right. (farmer 2, int. 3).

Evaluation of experiments

Farmers assess trials continously throughout the growing period. They used the phrase 'it went well' or 'it didn't go well', referring to the results of a trial. The criteria used by them in order to consider whether a trial 'went well or not,' include the following.

- Agronomic aspects: e.g. foliage, flower quantity, quality of the fruits.
- Total production (yield): this represents once of the main criteria for them to say whether a crop 'did well or not'.

- Economic benefits: this refers to the amount of money the farmer gets when selling the produce. This is really the 'key' criterion to determine whether the crop 'did well or not', because even if the other criteria (e.g. foliage and fruit resistance) were all right, if the products do not have a good price in the market, then they conclude that the crop 'did not do well'. One farmer stated,

> ...I have planted the tomato 'bonanza rio grande'. That is a very good one. It grows very well in this area, but it is not good when you try to sell it. Nobody wants to buy it; people don't like it. I have tried twice with it, but it didn't do well since people didn't buy it. (farmer 6 int. 1).

When two variables are being evaluated in double experimentation and something goes wrong, the trial is repeated, but this time changing only one of the variables used previously. A farmer explained when asked how one knows what went wrong in a double experimentation,

> that is something very easy. Let me tell you. If you are, for example, cooking and you put some seasoning in the food, let's say pepper and tomatoes, and at the end of the food doesn't taste good, then you conclude that it was either the tomatoes, or the peppers. It is the same when you are making a trial, if the trial doesn't go right, you try again, but this time you keep the same amount of fertilizer and change the planting distance to repeat the trial on the same piece of land. This way you can see what happened. It is very important to repeat the trial on the same piece of land. Those are the kinds of things that we farmers know. (farmer 3, int. 3).

Conclusion, implications and recommendations

- Small-scale farmers design and conduct most of their experiments in a self-directed way. Experimentation represents a key component of their strategy for survival. That means that these experiments are performed by all small-scale farmers in developing countries.
- Farmers' experiments do not occur in a vacuum. They show, among other things, the comprehensive understanding farmers have about the interactions that exist among the diverse factors affecting the production process (such as the socio-cultural, political and economic context) (Rengifo-Vasquez, 1989).
- There is a need for agricultural researchers, extensionist and development workers in general to build upon the existing local knowledge (including experimentation) to design and implement sensitive, cost-effective, bottom-up interventions that truly 'work' (McCorkle, 1994; Scoones and Thompson, 1994; Warren, 1991; 1992).
- There is a need to assist farmers to enhance and improve their own experimentation process. In order for agricultural researchers, extensionists and development workers to engage in meaningful communication with farmers, it is necessary 'to enter into the world of farmers' ideas, values, representations and performances' (Scoones and Thompson, 1994: 29). This can be done in different ways. One way is to explore methodologies that take into account the complexity of the farmer's world (Quiroz, 1992). These methodologies and approaches should be developed in a collaborative and interactive way with the farmer so they can be used by the farmers themselves and then can be disseminated to other farmers and facilitators (Chambers, 1994; McCorkle, 1994).
- Although farmers' experiments in one region may have general characteristics

that are common to the experimentation implemented by farmers in another region, they may also have characteristics particular to that region (e.g. availability of resources). In other words, although there is not a universal set of farmers' experimentation characteristics, what is important is to be sensitive to the fact that farmers experiment and therefore it is necessary to develop field methodologies and management strategies that support farmers in these roles.

8. Indian farmers opt for ecological profits

VITHAL RAJAN and M. A. QAYUM

Abstract

A GROUP OF voluntary agencies, agricultural scientists, farmers, and the Department of Agriculture, Andhra Pradesh State, India, are collaborating to use traditional, non-pesticidal methods to control the red-headed hairy caterpillar (*Amsacta albistriga-Walker*), a voracious polyphagous pest that destroys several rainy season crops of poverty-stricken farmers in the semi-arid tropical region of the Deccan, India. Timely community group action, by co-ordinating lighting of bonfires in contiguous red soil areas brought down the moth population of the pest in early years. To save on scarce burning material, the group has innovated by using hundreds of light traps for protecting 5000 hectares of castor and other crops. The project has shown the benefit of community group action, and there is expected to be a spin-off effect for introducing similar ecological technologies. Local activists feel the project also raises larger possibilities for social cohesion, and the role of people's knowledge in development.

The present-day context and key issues

The Indian countryside and the people of India are beginning to face the consequences of massive environmental degradation brought about by the stripping of forest cover and consequent soil erosion. The worst has happened within the last three decades. Forest cover has now shrunk to less than 10 per cent of total geographical area. Soil erosion is also a massive problem with loss of top soil in some watersheds of the semi-arid regions being of the order of 100 tons per hectare per year.

Along with the depletion of natural resources, India faces the problem of the widening gap between the rich and the poor, and the inability of economic policies to pull at least half of the population above the poverty line. The Green Revolution strategies, while certainly being of great benefit for the larger farmers in resource-rich areas, have failed to make an impact on the bulk of the rural population, composed of agriculture labour, small and marginal farmers, and others, especially those living on resource-poor lands and cultivating dry-land crops.

The rural population faces an endemic lack of employment – in some cases only 150 days of employment per year are available, especially for women. The lack of employment in this context has also been aggravated by the use of modern technology and capital-intensive methods for cash-crop cultivation. In addition, poor families are stricken by lack of food, and endemic hunger and malnutrition among women and children is a common experience among the poorest sections.

Scientific efforts at helping raise production and productivity of small farmers in resource-poor areas by institutions – such as the International Crop Research Institute for the Semi-Arid Tropics (ICRISAT), the flagship of international agricultural research institutes, located in the Indian Deccan plateau, and the Indian national research programmes – have not yet yielded tangible results.

A main constraint to genuine development has been perceived as the lack of people's genuine participation in the process. The emphasis on such participation, which is now being stressed by government and development experts, has been followed by a desire to revitalize people's grassroot organizations, and start development planning processes from the bottom upwards. However, while these development directions are stated in theory, very little has been done in practice to build up strong people's associations or *sanghams*![1]

People's science

Discussions have been carried out by members of the voluntary movement and groups about the meaning of 'people's science.'[2]

A distinction was made between traditional knowledge of people, and the 'science for the people' concept which extends modern scientific ideas to the poor.

The issue was seen as a political one in India, and more emphasis was placed on the social dimension than on the technological aspects. Indian thinkers and grass-root activists wish to avoid an approach that could be termed as supporting second-rate technology 'to keep the people poor,' or to fall back on mysticism and feudal practices. However, people in the voluntary movement were willing to give a trial to ideas more applicable to real grass-root development than present high-capital, high-technology intensive, top-down processes. There was also an added ecological awareness and a desire to integrate humanity with nature. Received nineteenth century Western science perspectives were seen as inadequate for such a purpose.

The following set of parameters could help define the action values to be adopted by people's science projects.

- Low-cost technology and cultural practices within the reach of the people.
- Strategies capable of implementation using personal, local, community skills.
- Strategies utilizing easily available materials.
- Strategies based on people's knowledge for assured social and cultural acceptability.
- Strategies generating employment.
- Ecologically sustainable strategies protecting:
  - plant and natural diversity
  - processes that do not exploit people or nature
  - processes based on organic farming principles rather than on chemicals.
- Agriculturally sustainable strategies based on:
  - multi-vector activities of the farmer
  - soil regeneration
  - optimal use of water.
- Strategies permitting community group action for mutual economic benefit and social solidarity.
- Strategies that lead to self-provisioning of communities (which may be more practical under the present circumstances than the more far-reaching Gandhian concept of self-sufficiency, or the ideological concept of self-reliance).
- Strategies that increase people's capacities for long-term planning/resilience, and which help absorb economic shocks and cyclic fluctuations of nature.
- Strategies that are 'error friendly' – that is, in which errors could cause least damage.

Intellectual domination

However, the hierarchy of governmental organization, as well as the social distance of the élites from the masses, has led basically only to a top-down approach in any sector of development activity.

This is true even of the voluntary movement. Even 'Left' activists have not conceded the scientific capacity of ordinary people, and there is no strong acceptance that people have the skills, capacities and knowledge to develop their own communities. Marxist and capitalistic ideologies share a common belief in scientific progressivism. This basic assumption of intellectual inequality is paternalistic, and as long as farmers, rural women, and the poorer sections of the community are not accepted on an equal intellectual footing, as capable of imparting knowledge as well as receiving it, genuine people's participation and grass-root planning will merely remain a wish for the future.

The ascendency of Western science and university-based knowledge over other systems of thought, has been questioned by some Indian scholars. They point to the sociological, imperial and colonial roots of such science in nineteenth century European expansion.[3] Others have investigated the wealth and relevance of Indian scientific traditions before the period of colonialism.[4] Other scientific activists have launched programmes to take science to the people.[5] Laudable as these efforts are, we have yet to see on any appreciable scale the involvement of ordinary people in the grass-root planning and implementation of development strategies which utilise the wealth of people's own experience and cultural practices. A possibility, perhaps, exists to make a start in an environmental–agricultural area of activity. Ethno-biological research and other studies have shown the excellence of local experience in solving practical problems. It may be important to build up the confidence of the people to extend this experience, under voluntary agency support, and with government approval.

Integrated pest management

It is in the area of pest management that modern science has within the last decade come to review drastically many of its assumptions.[6] Uncontrollable pest attacks in North America have brought foreign scientists and entomologists to perceive that pests can be managed by:

- promoting diversity of the habitat, including farming practices of mixed cropping and inter-cropping
- ensuring the presence of natural control processes, by which nature herself acts through her many species that have symbiotic relationships with each other, to keep pest populations down
- strengthening cultural practices by which poor farmers around the world have learnt to lessen crop loss without damaging the environment or incurring the cost of pesticides.

Integrated pest management has come to be scientifically acceptable, placing the least emphasis on the use of pesticides, which are to be used only under specific conditions. In this world-wide scientific climate for environmental approaches, it is appropriate that we should think of making a break-through in this area, not only because success will be spectacular, but success would earn the support of the environmentalists and, more importantly, of farmers, who can produce better crops at lesser cost.[7]

Traditional community-action pest-management practices are beginning to teach scientists that in many instances farming communities have a better understanding of the pest problem than do scientists. Treating pests as a sort of negative common property resource (that is, as a factor to be dealt with commonly by all) is the easiest way of energizing community group action – for such traditions are still remembered by farmers who in this generation have been seduced by government agri-technology extension agencies to use expensive pesticides. In the Deccan, these have led to eco-imbalances, pest proliferation, and extensive crop damage, leaving agricultural scientists with nothing to recommend except a return to past practices.

The red-headed hairy caterpillar (RHC) and the castor crop

The Telangana region of Andhra Pradesh State, particularly the districts of Ranga Reddy, Nalgonda, Warangal and Mahabubnagar, is a centre for castor growing in India. The kharif castor crop in Andhra Pradesh State is sown over 300 000 hectares, covering 67 per cent of all the land sown under this crop in India. However, yields of castor, which is typically a dry crop, sown on light red soil by small farmers, is as low as 170 kg/ha in A.P., accounting for only 18 per cent of total all-India production. Yields have gone down over the last decade because of the emergence of the red-headed hairy caterpillar (*Amsacta Albistriga Walker*) as a major pest. Entomologists believe that the pest's populations have increased because of biotic disturbance caused by the excessive use of pesticides, which have killed all their natural predators and other species of insects that used to occupy eco-niches. The pest is polyphagous, feeding on a wide varieties of crops, from sorghum to greengram and groundnut. It appears at the time when young castor shoots are in the fields. The pest destroys castor crops in the short period before it pupates and remains dormant for 10 months. While other crops recover, the earlier-sown crop of castor is totally destroyed and farmers are forced to sow again. Farmers have also been forced to delay castor sowing by a month or more, leading to the castor crop withering under water stress in its seedling stage (which occurs for late-sown castor in the post-rainy period).

Continuous failures of castor crops have led to despondency among farmers, who have lost control over local production of good quality seed, and other good farming practices. Gujarat farmers now dominate production in Andhra Pradesh State by being the main suppliers of seed. Marketing functions in Andhra Pradesh State are also controlled by non-local traders.

Dr. N. K. Sanghi, Regional Co-ordinator for the Transfer of Technology, Indian Council for Agricultural Research (ICAR), and a dynamic and committed scientist, helped to develop the concept of controlling the pest by using cultural practices that were once followed in different regions in South India. His ideas were supplemented by the field experience gained by the Jan Seva Mandal, a voluntary agency working in the Dhulia district of Maharastra.[8] When interviewed, Mr. Malakondiah, a retired Additional Director of Agriculture, Andhra Pradesh State, said he had used some of these methods very effectively against the pest many years ago in the Rayalseema region of Andhra Pradesh. The work of Dr. John Sudheer, entomologist working in Ananthapur District, A.P., also confirmed expectations that one could develop a viable non-pesticidal approach towards pest control.

With this kind of intellectual support, a programme was built up for the 1989 '*kharif*' or rainy season involving 10 local voluntary agencies in the districts of Ranga Reddy, Mahabubnagar, Nalgonda and Warangal. By 1992, eight agencies, with some changes, continued to be involved. Action for World Solidarity (ASW), a German donor agency, co-ordinated the efforts of these voluntary agencies. By 1992, OXFAM joined ASW in helping to fund and co-ordinate the project. Several meetings were held to understand the simple people-based technologies that had been suggested, and to organize the efforts of the agencies and the farmers to implement these technologies for pest control. Scientists under Dr. Sanghi's leadership co-ordinated training activities. The State Department of Agriculture not only supported the programme but gave about Rs.400 000 for providing burning material for lighting bonfires to destroy the moth – the most vulnerable phase of the pest. Support from the National Bank for Agricultural and Rural Development (NABARD) was sought and given. Under NABARD's leadership, commercial banks in the region agreed to extend credit to local farmers for agricultural inputs. However, in the event, local bank managers proved too conservative to be of real help to the poor. Starting with a simple idea, the approach involving several such agencies, and a total sum of around two million rupees was managed successfully by this *ad hoc* consortium of voluntary agencies. Such collaboration is unique for other than crisis situations. The reasons for successful co-operation could be listed as follows.

- Heightened environmental awareness among all voluntary groups, scientists and government officials.
- The sensitive and capable handling of issues, differences of opinion, and personalities; and the maintenance of a sense of independence and equality among the members of the group, particularly by ASW, the co-ordinating agency, Dr. Sanghi and Dr. Rahiman, the then Additional Director of Agriculture.
- The excitement caused by the possibility of effective grassroots collaboration between voluntary agencies and the government in a new unexplored area.
- The sharp focus of the project in time and space, and its delimitation around one pest (RHC) and one crop (castor).

Early results

The pest has four life stages: pupa, moth, egg and larva. It is a pest when it is a larva, feeding on crops as it grows in size till it pupates and remains dormant from August until the end of May. Rainfall triggers the emergence of the adult moth from the earth, but only visual sighting of large numbers of moths around lights can confirm this.

The technologies advised included summer ploughing to kill some pupae in the first few inches of the soil; followed by bonfires after monsoon rains in every field to kill moths before they mated; the collection of egg masses; and the planting of trap crops for attracting the caterpillars.

Bonfires must be lit in every field on the nights of mass emergence to attract and kill the moths before they mate, which they do within 48 hours. The egg masses (200 to 2000 laid in a cluster by every female moth) hatch in another 72 hours, and must be collected and destroyed before the tiny larvae emerge. The trap crops must already be in leaf at this stage to attract the travelling caterpillars. None of this can be done usefully by only a few farmers, but must be taken up as a village group activity.

While these methods saved on the use of pesticides (which were known to be ineffective in any case), the technologies had not been practised before, either by the farmers or by the voluntary agencies and the scientists. All four technologies were never practised in any one target area. Egg collection, which had proved useful in Maharastra, was not practised at all. Timing was essential, and most of all there was a need to co-ordinate the lighting of bonfires on the same night as the peak emergence of the moth. Many farmers, particularly in the villages around Jangaon, Warangal District, were convinced that the bonfires were effective in controlling the RHC pest, and were eager to continue the practice. Trap crops also needed to be planted earlier with the first crops. In a few places, the farmers were not able to achieve the co-ordination, timing, or skill in follow-up, needed to achieve results. However, in around 2000 hectares, castor yields went up from 170 kg/ha to between 500 and 600 kg/ha. A discussion with Mr. Lingaiah, working in the Jangaon area of Warangal District, revealed that the farmers not only lit bonfires in time, but also practised summer ploughing. Crops were reported to be badly affected in one village over eight hectares where bonfires were delayed. Effective timing and co-ordination was achieved by alerting farmers by beating a drum when hundreds of moths were sighted in street lights. It must be noted that farmers in the region were accustomed to taking ecological measures. They have used the unique practice in some villages of the region of attracting pest-eating birds by laying out cooked rice in the fields. Farmyard manure and silt at 125 cart-loads per hectare, costing Rs.250/ha, were also used. Mr. Lingaiah also experimented with growing 'Aruna' foundation castor seed, during the rainy season. An additional Rs.250 000 was produced for 150 farmers of Vadecherla village from 200 hectares. This clear success has encouraged all the partners in this experiment to continue.

Village seed-production, particularly of open-pollinated crops, such as castor, requires group action. Such crops are vulnerable to contamination from inferior strains, pollen of which can be carried by the wind. Hence the involvement of all the farmers growing such crops is necessary for a better variety of the crop to be grown and sustained in a region. In the experiment at Warangal district mentioned above, 'Aruna', a preferred local variety, was distributed to farmers who all agreed to grow this variety. Needless to say, control of seed and seed development at the village level will also help farmers maintain independence at the local level from business interests and large corporations.[9]

By 1992, six villages had begun growing quality seed which is supplied to all farmers of the sub-region. The State Seed Corporation has offered a premium of Rs.2.50 to Rs.3.00 per kg of farmer-produced seed over and above the market price. This is a testimony to the competence of the farmers.

Group action technologies

Dr. N. K. Sanghi has termed such people's science initiatives as Group-Action Technologies – because these are technologies that can be initiated only by the whole of the community taking joint action together. Such a definition is important to distinguish between people's science technologies and more conventional modern scientific methods, which can be applied by individual farmers irrespective of social commitment or the actions of the rest of the community.

These people's science experiments have thrown up many interesting research issues that have not so far been studied in research stations. While these instances may show the capacity of the farmers for close scientific observation and the ability to follow a scientific method, the key point to be made here is that those strategies work best for the poor that are done *on a community basis.* Pest management methods could not possibly succeed on an individual farm basis, and have to be adopted and co-ordinated by the whole community. In the identification of survival strategies or people's knowledge, then, community action or social process would take precedence over technological processes. None of these technologies could be usefully redesigned without considering the social relations and circumstances in which they were found useful. Or, in other words, social cohesion could also be considered to be the most essential element of such people-based knowledge.There are no neat ideas that can be lifted out of social context and capitalized, as, for example, some herbal medicines have been.

Recent developments

Most groups or agencies that search for alternative methods of development have evolved their approach based on certain ideological foundations.[10] There are resistances to working with the corporate sector; utilizing foreign money; employing modern technology; or co-operating with government officials. While the commitment of the voluntary agencies involved in this project is exemplary, and there is a genuine attempt to seek new directions, no insistence has been laid by any person or group on maintaining 'ideological purity'. This is another interesting feature of this project, which is conceived of and implemented in an open-ended manner. For example, both voluntary agencies and the government are experimenting in establishing electric-powered light traps.[11] Although this is modern technology, at present used by scientists for monitoring insect flight behaviour, it could rightly become a tool of people's science in the present-day context. In fact, it could be preferable to burning up scarce fuel-wood or toxic material such as rubber waste, which is the only other material available at present. Large-sized electric light traps could not only function better but save on time, which is also a priority in present-day rural India.

The availability of sufficient burning material for bonfires was a problem. An imaginative innovation in the 1992 programme has been the massive substitution of light traps for bonfires. Light traps protected over 5000 hectares during the 1992 *kharif*, or rainy season. Probably such a concentrated use of light traps had never been attempted before. As the evaluation report showed, this innovation has also proved successful. It is important to note that while light traps constitute an advanced technology, they are perhaps more appropriate since they can be managed by the farming community, and are more environmentally friendly than bonfires in an area already suffering from loss of vegetation and green cover. As pointed out in the report, the pest causes large-scale economic losses in castor crops covering an area of 80 000 hectares to 100 000 hectares in the Telangana region. When farmers do late sowing of castor to avoid the period of maximum pest infestation, they lose up to 20 000 tonnes of castor, valued at Rs.120 to Rs.140 million. The loss of other crops due to this pest can be accounted at another Rs.100 to Rs.150 million. Hence an additional value of a minimum of Rs.300 million worth of crops can be added if the vulnerable area is protected with light traps and vegetative traps, as this three-year programme has amply demonstrated. It is now expected that the Government, supported by voluntary agencies, will launch a large-scale training and implementation programme to bring this voracious pest under adequate management by utilizing a proven non-chemical technology.[12]

Preliminary assesments indicate that:

- light traps could be more 'appropriate' than bonfires

- o more emphasis should be laid on accepted cultural practices of manually destroying pests through group action
- o perennial vegetative traps should be laid
- o more agriculturists should be identified to lead and motivate farmers
- o traditional practices, such as offering cooked rice to attract birds to large caterpillar masses should be revived
- o scientists should collaborate with farmers to establish the range and efficiencies of different types of light-traps
- o above all, the programme should aim to empower farmers so that it can be 'owned' and managed by them. Only then, would there be a chance that such group action will be voluntarily replicated by other groups of farmers.

Future programmes

It had always been the intention of this group, involving research scientists and voluntary agencies, to broaden out from the experience of working with one pest and one crop, to start initiating group action at the village level, which will not only strengthen communities but also increase production and protect the environment. Within the next couple of years other projects that might be taken up include restoration of some village-tank irrigation systems, which only a hundred years ago numbered over 100 000 in this part of India.[13] Another aspect of group-action technology is community credit management, involving exchanges and repayments in kind. The Deccan Development Society, working in 60 villages of Medak District, has established *sanghams*, or associations of poor village women. They now control a credit pool of a million rupees. These village women have almost eliminated bonded labour in their villages; and by offering opportunities for co-operative farming have increased employment to 250 days per year per agricultural labourer in their villages. Wage rates have also increased.

Another project would be 'social fencing' in agro-forestry projects. That is, the community would take the responsibility for protecting the trees that it plants, and dependence for success would be shifted from expensive technical solutions, such as wire fencing, to more reliable social methods of control. This practice has already been established by some of the *sanghams* working with the Deccan Development Society. Another area of activity would be the introduction of dry-sown paddy in the *kharif* or rainy season, using a seed drill, followed by rabi or dry-season groundnut on residual moisture. This method extends the area under paddy by over 30 per cent, inhibits brown plant-hopper attack, facilitates more socially equitable distribution of tank water released later in September, fixes nitrogen in the soil through the action of the legume crop, and finally gives a double income to the farmer! Such water-saving methods using the traditional implement of the seed drill are now being widely practised over 20 000 hectares near Atmakur, Kurnool District, Andhra Pradesh.[14] The replacement of rabi or post-rainy paddy by irrigated dry crops is another possibility. Soil conservation methods involving field bunding to control rill erosion is another ecologically sound practice that requires community group action.[15] Agricultural scientists are already establishing the value of traditional multi-cropping and inter-cropping practices for pest management and sustainability. Dr. N. S. Jodha surmises that dryland farmers in the Telangana region practise around a hundred combinations of crops to hedge against agro-climatic variations.[16] The historic Ain-I-Akbari records of the Mogul Empire mention how several hundred years ago farmers in a particular region would grow as many as 40 taxable crops in a year![17]

A most important scientific question is whether this kind of work opens up the possibility of scientific research taking place which integrates ordinary farmers in participatory research. At present ICAR scientists cannot in general undertake activities called extension work. Nor can the Department of Agriculture extend technologies not approved by the ICAR. Under these boundary conditions existing between agricultural research and extension, projects such as the above cannot be carried out except under the auspices of voluntary agencies. It is this form of scientific research demarcation which has given the project a unique collaborative quality.

Within the next few years the group expects to initiate experimental projects on a small scale involving people's participation and group action technologies to promote sustainable farming practices that take care of the earth as well as the poor people who live in the region[18]. Such a people-based science would also be spiritual and social.[19] Since it can be implemented and controlled only at the grass-roots, this form of science would be a true extension of humanity.

9. Indigenous agricultural experimentation in home gardens of South India: conserving biological diversity and achieving nutritional security.

B. RAJASEKARAN

Abstract

INDIGENOUS AGRICULTURAL EXPERIMENTS are often conducted by farming communities with a broad objective to conserve biological diversity and achieve nutritional security. Farmers' and gardeners' knowledge of domesticated and wild species form the foundation for indigenous agricultural experimentation. The typology, biodiversity, and cultural practices adopted in various home garden systems are discussed. Specific cases are provided to illustrate the rationale behind conducting experiments in home gardens. Lessons learnt from farmer experimentation would provide essential insights for developing biotechnology programmes for increasing home garden productivity by public and private industries.

Introduction

Indigenous agricultural experimentation is the process by which local people, including farmers, gardeners and farm labourers, informally conduct trials or tests that can result in innovative farming techniques, cropping systems, and management practices suitable to their agroecological, sociocultural, and economic conditions. Indigenous experiments are often conducted on a trial-and-error basis by making use of the physical and biological resources available, such as local seeds, manures, land, and labour. More importantly, indigenous knowledge based on personal experience as well as that of one's elders, friends, and relatives acts as a basic input for conducting the experiments.

Indigenous agricultural knowledge of local people plays a crucial role in conserving biological diversity and achieving nutritional security (Altieri and Merrick, 1987; Rajasekaran and Whiteford, 1993; Rajasekaran and Warren, 1994; Richards, 1985; Warren, 1991; 1992). Farmers possess an in-depth knowledge of domesticated and wild species of plants maintained on farms and gardens. They are also familiar with related agroecological and sociocultural conditions that influence cultivation of these species. Home gardening, a type of food production based on small plots adjacent to human settlements, is the traditional and most enduring form of cultivation (Ninez, 1987). Home gardens have been a part of human subsistence strategies since the Neolithic Age (Cleveland and Soleri, 1987). They have played a meaningful role in the domestication of plants and continue to be an avenue for the introduction and adaptation of new crop varieties. Farmers regard the home garden as an association of individual plants or groups of plants, arranged according to empirical knowledge of the environmental needs of each plant in terms of shade, humidity, synergy, and competition with other crop species. Home gardeners are the custodians of indigenous knowledge related to diversity, interactions, cultural practices, and *in situ* conservation of the home garden species.

Recent ethnoecological, archeobotanical, and paleobotanical studies have indicated that indigenous management practices have influenced the present-day abundance of home garden species such as *Annona spp., Byrsonima spp., Carica spp., Ficus spp., Manilkara spp., Quercus spp.*, and *Spondias spp.* (Gomez-Pompa, 1987; Harlan, 1975; Hynes and Chase, 1982; Kundstadter, 1978; Posey, 1990; Roosevelt, 1990; Turner and Miksicek, 1984). Home gardens represent a 'genetic backstop', preserving species and varieties that are not economical in field production and are planted on a small scale for reasons of taste, preference, tradition, or availability of plant materials (Ninez, 1987). Of approximately 300 major vegetables favoured today the world over, 200 are produced by backyard gardeners, while only 20 are used in field cultivation (Grubben, 1977). Despite a number of new varieties of vegetables released from research stations in recent years, the following factors influence farmers to continue the adoption of traditional varieties: (1) farmers perceive that traditional varieties of vegetables are more delicious and nutritious (Rajasekaran, 1994); (2) traditional varieties are well adapted to marginal soils and adverse climatic conditions; (3) the seeds of newly released vegetable varieties may not be easily available to farmers and home gardeners, who also may not be able to afford to buy the new varieties; (4) cultivation techniques of traditional vegetables are known and are often easier; and (5) traditional varieties are often resistant to pests and diseases.

Methodology

Data for the study were collected from Kizhur and Pillayarkuppam villages of the Union Territory of Pondicherry, India. Through participant observations, typology and the recording of indigenous knowledge, biological diversity and cultural practices reflected in the management of home gardeners were explored. Experiments conducted by home gardeners on a trial-and-error basis were also recorded. Home gardeners were interviewed to explore the sociocultural, ecological, economic, and technological factors influencing the decisions involved in the selection and adoption of different types of home gardens.

Home gardens: typology, biodiversity, and cultural practices

Farms and human settlements are found in two distinct locations within the boundaries in the study villages. The distance between agricultural fields and farm habitats is usually less than half a kilometre. The home gardens are traditionally referred to as *thottam* whereas the agricultural farms where field crops are grown are referred to as *pannai* or *vayal*. Household gardens tend to be located close or adjacent to dwellings for security, convenience, and special care. They may occupy land marginal to field production and involve labour marginal to major household economic activities (Ninez, 1987). Home gardens are typically populated by a wide variety of plants, varying from small herbs to tall trees (Soemarwoto and Conway, 1991). The following principles of home gardeners encourage them to place high value on the home gardens: (1) consume what is produced in the garden; (2) reduce the dependency on outside vegetable markets; (3) consume fresh vegetables as far as possible; (4) use irrigation water sparingly; (5) make efficient use of space surrounded by or adjacent to the dwelling for gardening; and (6) use the garden products as a source of secondary income.

Based on certain physical and biological characteristics, home gardens in the Pondicherry region of India, can be classified into four types: (1) backyard gardens; (2) rooftop gardens; (3) upland gardens; and (4) pump house gardens.

Backyard gardens

Farmers' dwellings that are located within the boundaries of the farm often reserve space for backyard gardens. The farmers and backyard gardeners who build their houses on the farm are isolated from the larger human settlements. Living in settlements is not always conducive to garden maintenance, especially when considering problems such as crop thefts and cattle menace. It is important that at least one family member should be available in the vicinity of the garden during both the day and night to take care of the garden. The backyard gardeners give priority to their gardens rather than to maintaining their social lives, by living closer to the rest of the farming communities. However, they often visit their friends and relatives in the village settlements to maintain their social relationships.

A total family involvement was observed in backyard gardening. The backyard gardens exhibit a high division of labour and expertise. Older members of the families are involved in selecting seeds for sowing. Younger family members and women are responsible for garden activities and compost pit maintenance, respectively. The backyard gardens have a high degree of species diversity (Table 9.1). Seasonal herbaceous vegetable crops and fruit trees form the primary components of these gardens. Raised bed nurseries are formed for sowing small-seeded vegetable crops. Raised beds effectively control fungal diseases. Thurston (1991) provides a detailed discussion on raised bed nurseries and their impact on reducing the incidence of plant diseases. Backyard gardeners sow seeds of brinjal, chillies, and tomatoes in nurseries. After three to four weeks, well-established seedlings are pulled out and planted in the main field. Okras, radishes, and cluster beans are sown directly in the main field by forming ridges and furrows.

The compost pits play a major role in supplying nutrition to backyard gardens. At least one compost pit is found in each of the gardens. The women collect dried leaves of trees, cow dung, goat dung, paddy straw, and other dried plant materials, and dump them into the pit. Upon filling, the pit is closed air tight for approximately two months. Due to anaerobic decomposition, the leaves and other organic materials decompose. The organic manure is applied mainly as basal dressing just after field preparation. The compost manures are cost-effective as they substitute for chemical fertilizers. Though comparatively slow in action, they are environmentally friendly and replenish soil fertility. The backyard gardeners also rear country chickens, with the excreta applied as manure to fruit trees.

Fruit trees such as banana, mango, guava, lemon, and jack are grown with other leafy vegetable trees such as *moringa* and *sesbania*. In the intermittent spaces of trees, tuber crops such as *karunai kizangu* and *sepan kizangu* are grown. Coconut trees are planted near the field boundaries of the backyard gardens. Fodder trees such as *leucaena* also form part of the live fence of the backyard gardens.

Women play a major role in home garden production. Apart from filling the compost pits, they are also involved in harvesting vegetables for home consumption, protecting the gardens from cattle menace and crop thefts, and preparing preserved foods such as mango pickles, lemon pickles, and lime pickles.

Backyard gardeners are also producers and suppliers of vegetable seeds of brinjal, tomato, okra, chillies, cluster beans, flat beans, and gourds. The rest of the farming communities depend to a significant extent on these seed producers. During cultivation, the gardeners leave one or two fruits on the plant itself for

Table 9.1. Species diversity of home garden species and their utilization

Common name	Local name (in Tamil)	Botanical name	Utilization				
			Vegetable	Fruit	Pickles/ dried food	Fodder	Firewood
Herbaceous crops:							
Eggplant (3)	*Kathiri*	*Solanum melongena*	*				
Tomato (1)	*Thakkali*	*Lycopersican esculentum*	*	*	*		
Okra (3)	*Bhendi*	*Hibiscus esculentus*	*				*
Radish (1)	*Mullangi*	*Raphanus sativus*	*				
Pepper (1,3)	*Milagai*	*Capsicum annum*	*		*		*
Cluster beans (1)	*Beans*	*Cyamposis tetragonobus*	*				
Peas (1)	*Pattani*	*Pisum sativum*	*				
Amaranth (4)	*Arakeerai*	*Amaranthus tricolor*	*				
Taro (1)	*Senai*	*Colocacia esculenta*	*				
Purslane (4)	*Gonkra*	*Portulaca oleracea*	*			*	
Tree crops:							
Sapota (1)	*Sapota*	*Achras sapota*		*			
Annona (1,4)	*Seetha*	*Annona muricata*		*			
Guava (1,4)	*Goiya*	*Psidium guajava*		*			
Banana (1,4)	*Vazhai*	*Musa Sp.*		*			*
Cowpea (1)	*Karamani*	*Vigna Sp.*	*				
Coconut (1,2,3)	*Thennai*	*Cocos nucifera*	*				*
Coriander (3)	*Kothamalli*	*Coriandrum sativum*	*			*	
	Gonkra	*Portulaca oleracea*	*			*	
Anacardium (3)	*Mathulai*	*Anacardium occidentale*		*			
Sesbania (4)	*Agathi*	*Sesbania grandiflora*	*				
Lemon (1)		*Citrus aurantifolia*		*	*		
Lime (1)		*Citrus aurantium*		*	*		
Moringa (1,4)	*Murungai*	*Moringa oleifera*	*			*	*
Mango (1,4)	*Ma*	*Mangifera indica*	*	*	*	*	

Table 9.1. Species diversity of home garden species and their utilization (continued)

Common name	Local name (in Tamil)	Botanical name	Utilization				
			Vegetable	Fruit	Pickles/ dried food	Fodder	Firewood
Live fences:							
Unknown (1,4)	*Sundaikai*	*Solanum nigrum*	*				
Cassava (3)	*Maravalli*	*Manioc utilisima*	*			*	*
Fig (1,4)	*Athi*	*Ficus carica*	*				
Papaya (1,4)	*Papali*	*Carica papaya*		*			
Tamarind (1)	*Puli*	*Tamarindus indica*		*		*	*
Climbers:							
Lablab (1)	*Avarai*	*Dolichos lablab*	*				
Bitter gourd (2)	*Pagal*	*Momordica charantia*	*				
Snake gourd (2)	*Pudalai*	*Tricosanthes anguina*	*				
Bottle gourd (2)	*Surai*	*Lagenaria siceraria*	*				
Sponge gourd (2)	*Peerkai*	*Luffa cylindrica*	*				
Pumpkin (2)	*Poosani*	*Curcurbita Spp.*	*				
Muskmelon (2)	*Moolam*	*Cucumis melo*		*			
Indian spinach (4)	*Pasalai heerai*	*Basella alba*	*				

1–Backyard garden crops
2–Rooftop garden crops
3–Upland garden crops
4–Pump house garden crops

seed purposes. Women possess a wide range of knowledge regarding vegetable seed production such as selection of seed plants, selection of fruits, time of harvesting, separation of seeds from flesh, seed drying, and seed preservation techniques. They claim that the seed processing and preservation techniques contribute significantly to seed genetic viability and germination capacity. Women store the seeds and make them available to men whenever they ask for them. Generally, the seeds are stored in earthen pots since they provide an optimum temperature for maintaining the genetic viability of seeds.

Irrigation water is lifted from the ground by motor pumps. Most of the herbaceous crops are seasonal in nature and hence they require water almost daily. Most of the produce from backyard gardens is used for home consumption, with any surplus being disposed of informally in the village.

Rooftop gardens

Landless agricultural labourers grow gardens on the rooftops of their dwellings. Lack of sufficient garden space and the availability of thatched roofs have induced this group of villagers to innovate the practice of rooftop gardening. The walls of the huts are constructed with bricks and clay. The roofs are thatched with palm fronds in such a way to obtain a slope of 25 to 30 degrees. The palm frond roofs are supported underneath by bamboo poles. These types of houses are generally found in the South Indian villages and are referred to as huts (*gudisai*). Inside the huts, clay walls are raised to divide the floor area into rooms. The area of a hut ranges from 200 to 300 square feet.

Taking advantage of the slope of the roofs, these gardeners grow gourd vegetable crops. The gourds are climbers and hence are well-suited for rooftops. The gardeners sow seeds of gourd vegetables such as bottle gourd and sponge gourd, and pumpkins in pits around the base of the hut about 12" to 18" deep. The soil is loosened and then the pits are spread with farmyard manure. This is followed by sowing the seeds in the pits. In 4 to 6 weeks, the gourd plants give rise to vines and the gardeners use poles of casuarina trees to lift the vines. The casuarina poles are laid in a slanting position in such a way that the tips of the poles touch the lower end of the roofs. This arrangement enables the vines of the gourds to reach the roofs. Within two months the vines of the gourds spread over the entire rooftops. This is followed by flowering and fruit formation of the gourds.

Rooftop gardens are less labour-intensive since they do not require continuous care as in the case of other types of gardens. The rooftop gardeners irrigate water in the root zone on alternate days during the first month, and two days a week after the first month. Pest infestation is minimal and no weeding is required. The men climb on the rooftops to harvest the gourds. At frequent intervals, they also scout for insect pests and diseases. The rooftop gardens also provide a cooling effect to the houses during the summer months.

The pumpkins fetch a good price at farm markets. The Gloria Farm located in Pillayarkuppam village serves as the market outlet for the pumpkins. The Gloria Farm purchases the pumpkins from the rooftop gardeners and transports them to Aurobindo Ashram Food Service Center located in the city of Pondicherry. Hence, the rooftop gardeners have a steady demand for their produce. However, limited space and seasonal production of gourds restrict them from meeting the market demand. The rooftop gardeners reserve gourd vegetables such as the sponge gourd and bottle gourd for home consumption. These gardens contribute significantly to off-farm income and the nutritional security of the landless agricultural labourers. Rooftop gardens were also observed by Cleveland and Soleri (1987). In West African savannahs, during the rainy season, okras, amaranth, and other vegetables are grown by women in small garden plots adjacent to the houses, with vines such as luffa and pumpkin growing over rooftops.

Upland gardens

In Kizhur village of Pondicherry region, upland gardens complement household economies based on subsistence/cash crop production. Hence, these types of gardens could also be referred to as market gardens. Young farmers plant chillies, tapioca, and brinjal (egg plant) on irrigated upland fields. The upland gardens are generally located in slightly elevated areas that are found between crop fields and human settlements. Households receive no nutritional benefits from the upland gardens. These gardens are managed by young farmers. They allocate 5–10 cents to each one of these crops: chillies, tapioca, and brinjal. Mostly, the chilli gardens are managed by young farmers. The chilli gardens are labour-intensive and hence are not preferred by older farmers.

As a result of the high demand for green chillies in the vegetable markets, the entrepreneurial young farmers have introduced the cultivation of chillies, however, with the support of their families. Raising nurseries and sowing, preparing the main field and planting, irrigating, pest controlling, weeding, and harvesting make the upland gardens more labour-intensive. The sandy loam soils are well suited for chilli gardens. By creating an informal network, young farmers exchange the seeds among themselves. They claim that the practice helps to maintain the production of chillies. In other words, using the same type of seeds year after year results in poor yield performance.

Due to their multiple uses, such as vegetables, spices, and in condiments and pickles, chillies are in high demand in the market. Two types of marketing processes were observed in the upland gardens. Tapioca is sold in the field itself to merchants, loaded on to trucks and taken to town markets. Using transport buses, the farmers take chillies to market and sell them to vegetable vendors. Although labour-intensive, upland gardeners feel that gardening is a remunerative enterprise. It is interesting to observe a high diversity of species in the chilli gardens. One young farmer grew four different varieties of chillies in his garden. By frequently exchanging and introducing new chilli varieties, the young farmers maintain the diversity and productivity of chillies.

Pump house gardens

Pump houses are located at one end of the farms where the irrigation motor engines are installed. Ground water that is drawn using the pump engines cater to the irrigation needs of major field crops such as rice, finger millets, groundnuts, and sugarcane. The pump houses, with a 10 x 8 foot area, are surrounded by trees and shrubs. This location also serves as a resting place for farmers and farm labourers. Herbaceous leafy vegetables and tree crops are cultivated near the pump houses. After irrigating the major crops, the farmers use a small portion of irrigation water for the pump house garden crops. In addition, excess water found in the rice fields is also drained to these gardens. This demonstrates how farmers use their irrigation water efficiently.

Fruit trees such as mango, papaya, coconut, fig, and guava are grown in the hedges of these gardens. No strict labour allocation has been observed in this type of garden. However, women concentrate on planting, irrigating, and harvesting green leafy vegetables such as *thandu keerai, siru keerai, arai keerai, mulai keerai,* and *mullangi.* The rationale behind the use of a wide variety of green leafy vegetables in the pump house gardens is that the tastes of these vegetables are already familiar to and accepted by the local people. They also require continual irrigation to maintain their freshness and succulence (Mwajumwa *et al.*, 1991).

The women adopt different methods of harvesting the leafy vegetables. They completely uproot *thandu keerai, siru keerai,* and *mulai keerai* just before flowering. Once these crops start flowering, the green leaves become unfit for home consumption. In the case of *pasalai keerai, karisilan kanni,* and *manathakkali,* the women cut secondary and tertiary leaves, leaving the apex and tender shoots for regeneration. The apex and tender shoots give rise to leaves in two to three weeks, and the harvesting cycle thus continues. Coconut trees and bananas are found in most of the pump house gardens. The leafy vegetables and fruits complement household consumption.

The type of farmers varies from garden to garden (Table 9.2). For instance, landless labourers are involved in the rooftop gardens, whereas young farmers play a pivotal role in the upland gardens. The intensity of labour and the level of women's involvement also vary among the various types of gardens (Table 9.2). Upland gardens tend to be input-intensive whereas the other three types of gardens use local resources such as local seeds and organic manures sparingly. Of the four types of home gardens, backyard gardens exhibit a high density of species diversity where as upland gardens exhibit varietal diversity of a particular species (Table 9.1).

Farmer experimentation in home gardens

Home gardeners are guided by their own experiments and perceptions about species selection and horticultural management practices so that each garden is an experimental unit itself. Gardeners conduct experiments in order to test their ideas and beliefs in their own way. Hence, as Roling and Engel (1992: 127) observed, 'to look at farmers only as users neglects the important fact that farmers are experimenters and that farmers have developed most of the innovations that are used on the farm today'.

The following factors influence home gardeners in conducting experiments in home gardens: (1) Marketing: the demand for home garden vegetables (chillies) in Pondicherry town markets encourage farmers to test different varieties of the same garden crop (e.g., green chillies); (2) Temporal: availability of sufficient time induces farmers to observe the performance of different varieties of species under varying conditions of gardens. For instance, by constant observation rooftop gardeners found that sponge gourds perform well when lifted by a neem tree; (3) Ecological: farmers are interested in identifying which herbaceous vegetable crops perform well when planted adjacent to which trees. For instance, backyard gardeners found that taro, a tuber vegetable crop, performs well under partial shade of *agathi* (*Sesbania sesban*), a shrub; (4) Spatial: lack of space for gardening forces landless labourers to test gourd vegetables on the rooftops of their huts; (5) Economic: lack of financial resources restrict backyard gardeners from purchasing seeds from government depots and seed markets. This situation stimulates them to experiment constantly with vegetable seeds received from their friends, relatives, and neighbours.

Table 9.2. Sociocultural, technological, ecological, and economical parameters of home gardens

Parameters	*Backyard gardens*	*Rooftop gardens*	*Upland gardens*	*Pump house gardens*
Sociocultural:				
Farmer type	Marginal farmers	Landless labourers	Young farmers	Farm women, marginal farmers, small-scale farmers.
Labour intensity	Highly intensive	Not intensive	Highly intensive	Moderately intensive.
Gender role	Moderate involvement of women.	No women involvement.	No women involvement.	Women play a major role.
Interaction among producers	Interaction for seed exchange.	Interaction for seed exchange.	Interaction for marketing.	No interaction.
Caste	Backward caste.	Scheduled caste.	Backward caste.	Forward caste, backward caste.
Technological:				
Demand for quality seeds	Moderate demand.	No demand	High demand	Moderate demand.
Level of mechanization	Hand tools to mechanized.	Hand tools	Hand tools to mechanized.	Hand tools
Demand for fertilizers	Low demand	No demand	High demand	Moderate demand.
Incidence of pests and diseases	Moderate incidence.	Low incidence	High incidence	Low incidence.
Harvest frequency	Weekly, seasonal.	Seasonal	Seasonal	Weekly, seasonal.
Ecological:				
Species diversity	High	Medium	Medium to low	Medium to low.
Species type	Herbaceous, shrubs, trees.	Vines	Herbaceous	Herbaceous, shrubs, trees.
Location	Adjacent to homes.	Roof of homes	Away from homes.	Close to homes.
Space utilization	Horizontal, vertical.	Horizontal, vertical.	Vertical	Horizontal, vertical.
Cropping pattern	Irregular, row.	Irregular.	Row.	Irregular, row.
Soil type	Sandy loam	Clayey	Sandy loam	Sandy loam, Clayey.
Economical:				
Production objective	Home consumption.	Market, home consumption.	Market	Home consumption.
Cost of cultivation	Medium.	Low.	High.	Medium.
Marketing infrastructure	Not applicable	Marketing support available.	Availability of marketing co-operative.	Not available
Extension services	Frequently available.	Not available	Frequently available.	Rarely available.
Credit support	No credit support.	No credit support.	No credit support.	No credit support.

Space utility experiments

Rooftop gardeners conduct experiments with the objective of utilizing efficiently the limited space. Depending on the availability of space, rooftop gardeners alter their sowing techniques. A few rooftop gardeners have sown different varieties of gourd vegetables in a single circular pit, whereas others dug separate pits for each variety of gourd. The size of circular pits also varies from 12" to 18" in diameter. In another case, a rooftop gardener lifted the bottle gourd vines on a pongamia tree, sponge gourd vines on a neem tree, and pumpkin vines on the rooftops of his house. The gardener claims that using different types of supporters to lift various gourd vines leads to lessening the competition among the gourds, and reduces the crowding of vines and thus results in easy harvest and increased yields, with minimal pest and disease incidence.

Market testing

Young farmers of upland gardens are profit-oriented and are interested in analysing the market demand for garden-grown vegetables. They constantly watch the demand for chillies in the market and adjust the picking of vegetables accordingly. The young farmers have joined together as an informal marketing society wherein they brought the merchants to their door, thus eliminating the middle-men. Conventionally, the farmers have to take their produce to the markets or *shandies* in towns. Constant market analysis and co-operative skills have also turned the upland gardens into an economic enterprise.

Crop combination testing

Backyard gardeners constantly experiment with crop rotation and crop combination techniques. Through experimentation, they found various combinations of herbaceous and tree crops. They also experiment with shade-loving and shade-hesitant plants. Elderly women display a remarkable array of knowledge of crop combination techniques in the pump house gardens. When compared with other types of home gardens, the pump houses are prone to more shade. Hence, repeated informal testing and experimentation made the elderly women familiar with shade-loving leafy vegetable crops. For instance, *karisilankanni*, a green leafy vegetable crop, performs well under shady conditions due to the rough leaf surface. Pump house gardens are also sites of poorly drained soils. Women tested different types of banana and coconut in poorly drained soils. One of the women farmers claims that *poovan*, a banana variety, performs well under the waterlogged conditions of the pump house gardens.

Varietal testing

The young farmers conduct experiments by constantly changing the varieties of chillies. Informally, they visit each other's chilli gardens and select seeds. They make identification marks by tying a thread near the selected fruit. According to some farmers, changing the seed varieties season after season increases the yield.

In a pump house garden, about six different varieties of banana were tested. Such varietal testing was also observed elsewhere. In one river basin in West Java, the performance of 34 banana varieties were recorded (Soemarwoto and Soemarwoto, 1984). The fruit of some bananas are eaten as dessert, or steamed for snacks, and others are supplementary staples. Home gardeners also clearly recognize the long-term importance of genetic diversity. When asked why an unused tree is found in a garden, they typically respond by saying that they might need it sometime in the future (Soemarwoto and Conway, 1991).

Informal IPM experiments

Home gardeners are meticulous IPM (Integrated Pest Management) experimenters. Farmers experiment with different types of pesticides, especially in the backyard gardens. They found that dimethoate, an insecticide, is highly effective in controlling the insect borers that infest the fruits of brinjal and okra. If the borer incidence is severe, they normally skip the planting of that vegetable crop for the successive season. To control minor incidence of pests and diseases, home gardeners apply cow dung ash. During sowing, elderly people are involved in reflecting the sunlight using mirrors to scare away crows to prevent the birds from carrying away the seeds. Since the small seeds of brinjal, okra, and other leafy vegetables are sown on the soil surface, protecting the seeds from birds and ants is crucial to maintain the required plant population. Two or three days before planting, they check for ant burrows in the garden and apply kerosene to the burrows in order to drive the ants away from the garden.

It is clear that, through experimentation, farmers demonstrate abilities to cope with and adapt to change, to confront contemporary and future situations, and to reassess existing adaptive strategies to deal with environmental and socioeconomic forces (Dei, 1988; 1989). Farmers, with their constant interest in experimentation, have innovated a wide spectrum of home garden technologies that are ecologically sound and socioeconomically benign.

Nutritional security in home gardens

Most of the value of the home garden to the rural household lies not so much in net income generated but in the range of production and its contribution to the overall livelihood of the household (Soemarwoto and Conway, 1991). Maintenance of continuous food production throughout the year is one of the important roles of the home gardens. In general, there is always something to harvest from the home gardens. Though the economic status of people who depend on home gardens is lower than for large-scale farmers, their accessibility to fresh garden vegetables is relatively higher. In other words, they are nutritionally more secure than some farmers who depend exclusively on monocropping and cash cropping.

Local varieties of gourd vegetables, such as bitter gourds, snake gourds, and ribbed gourds, grown by marginal farmers in the coastal village of Shollinganallur, Tamil Nadu State, India, are found to contribute significantly to nutrition (Rajasekaran and Whiteford, 1993). Research done elsewhere in India showed that indigenous vegetable production significantly improved the nutritional status of weaning-age children (Kumar, 1978). In yet another study, low-income households consumed the least amount of rice, but the largest amount of leafy vegetables, harvested from Javanese home gardens (Soemarwoto and Conway, 1991). Vegetables obtained from home gardens are rich in vitamin A and vitamin C, and provide nearly 40 per cent of the household requirements for energy (Stoler, 1979). Some plants consumed for one reason, may, in fact, serve multiple purposes. In southern Mexico peasants garnish their rice and beans with small amounts of the herb *epazote (Chenopodium foetidum)* (Whiteford, 1995).

A comparative study conducted by Soemarwoto and Soemarwoto (1984) of the production and nutritional value of three predominant agricultural systems – home gardens, *taluk-kebun* (an agroforestry system in Java), and rice fields – demonstrated that their production levels did not vary greatly. However, for nutritional value, the home gardens and *taluk-kebun* were better sources of calcium, vitamin A, and vitamin C than were rice fields (Soemarwoto and Soemarwoto, 1984).

Women preserve vegetables and fruits for long-term usage by making pickles from mango, lime, lemon, and ginger. Dried foods (*vathal*) are prepared from vegetables such as bitter gourd and a solaneous vegetable known as *sundai*. According to women, most of the leafy vegetables have significant medicinal properties. For instance, leaves of *manathakali*, a leafy herbacous crop is traditionally used as a medicine to control mouth lesions and other vitamin A deficiencies. *Karisilankani* and *keezhanalli*, leafy vegetables, are found to be effective in controlling yellow fever.

Conclusions

Despite the increased awareness of the value of indigenous knowledge and recognizing farmers and gardeners as experimenters, current use of home gardeners' knowledge and experiments for research and development is minimal. Policy initiatives for policy-makers, researchers, and extensionists are necessary in order to incorporate the knowledge and innovations generated by these local-level researchers (Rajasekaran, 1994; Rajasekaran and Warren, 1994). Integrating home gardeners' experiments and formal horticultural research is essential in order to conserve biological diversity, achieve nutritional security, and increase home garden productivity.

Restructuring horticultural research and extension programmes by keeping the existing fund of knowledge in home gardening is crucial to the successful adoption and dissemination of formal horticultural research. Farmer experiments could provide essential data and information on biological diversity, species combination techniques, varietal adaptability, integrated pest management, and local accessibility, to the horticultural researchers. At the same time, formal science and technology could contribute significantly to home garden production through crop improvements using advances such as biotechnology. This is especially important when circumstances of home gardens change, for example, due to increased drought, climate change, decreased availability of organic matter, or when people move to a new environment (Cleveland and Soleri, 1987). Horticultural researchers should conduct on-station experiments and garden trials by appropriately integrating both indigenous and agronomic knowledge systems. There are a number of lessons that extension personnel could learn from these types of home gardens. The types of home gardens, factors influencing the maintenance of a particular type of home garden, and farmers' informal experimentation and innovation should be taken into consideration when disseminating horticultural technologies.

The horticultural experiment stations should make efforts to collect local species of home garden vegetables. Both *in situ* and *ex situ* methods need to be employed to conserve the biological diversity of home garden species. The local genetic materials could form the foundations for horticultural research programmes such as crop improvement for yield, pest and disease resistance, climate change, and environmental stress. During the process of collecting these invaluable local genetic materials, farmers need to be suitably rewarded. The local government should allocate funds for rewarding the gardeners who maintain and use the rare species. For instance, the gourd crops that are used in the rooftop gardens are partially drought-tolerant and possess hardy stems. These varieties could be used as resource materials for developing drought-tolerant gourd crops.

Crop-theft is one of the major social problems in the Pondicherry region. Although farmers in the upland areas of the study villages are interested in diversifying their crop production by including garden vegetables in their cropping systems, crop-theft acts as a great barrier to such diversification. Awareness of this problem and steps to prevent crop-theft are essential. Local organizations and NGOs could play a major role in empowering home gardeners in this process.

Strengthening the local marketing infrastructure is also essential for upland gardens whose survival depends on outside markets. Marketing societies for garden surpluses should be established, based on the model of existing tapioca marketing societies (Rajasekaran, 1994).

10. Living local knowledge for sustainable development

NIELS ROLING and JAN BROUWERS

Abstract

INDIGENOUS KNOWLEDGE IS often seen as an achievement of the past, to be conserved in the present. The gene bank, and *in situ* conservation of agro-biodiversity, are typical examples. This chapter goes further and looks at indigenous knowledge as a living resource that is constantly re-invented. Such re-invention can be actively facilitated. The chapter provides a number of examples, which deal with efforts to promote more sustainable forms of land use, be it at the farm or higher system levels (e.g., water catchment). Sustainable land use is more knowledge-intensive than conventional land use. Facilitating it requires the active creation of local knowledge through discovery learning and inter-subjective social learning.

Introduction

Indigenous knowledge (IK) is often seen as an achievement of the past, now becoming extinct under the impact of modern scientific knowledge. It has the attraction of a pre-industrial artifact. Consequently, the focus of IK research is often on preserving what is left. A typical example is the gene bank. Indigenous technologies and farming systems information can be stored in a similar manner in depositories and libraries, and much effort is being devoted to such conservation. In recent times, this effort has been expanded by paying deliberate attention to the local knowledge and procedures involved in using the technologies. It is of little use to store seeds of a specific crop variety, if one does not know the conditions and purposes for which local people developed and used it. A typical example is the Dinka woman who has six different sorghum varieties in her clay storage bin, each one with its own advantages and disadvantages with respect to cooking qualities, taste, waterlogging resistance, resistance against bird attack and so on. Without the software (the woman's knowledge), storing the hardware (the seeds) is not much use.

A further recent advance with respect to maintaining diversity is *in situ* storage, creating the conditions for local people to maintain local genetic diversity or specific technologies as a dynamic resource.

The present chapter goes further. It presents a perspective on the *creation* of living local knowledge (i.e., local knowing) as a condition for sustainable agriculture. It argues that facilitation of such knowing is a key strategy in the quest for more sustainable agriculture. Such a perspective implies that, instead of a past achievement in need of conservation for the present, local knowledge is a present condition for having a future.

The chapter starts with the evidence in the form of case-studies which we partly glean from the literature and partly from our own observation. We shall consider:

- the facilitation of local soil classification as a condition for conservation measures by landcare groups in Western Australia. This case-study is based on direct observation by Röling and on Campbell (1994), the previous national landcare facilitator

- 'farmer field schools' as venues for creating local knowledge about pests and their predators as a condition for successful IPM in Central Java. This case-study is based on Röling and Van de Fliert (1994) and Van de Fliert (1993). The latter reports on an evaluation of the farmer field schools
- a network of farmer experimenters in the Mono Province in Benin as a device for accelerating local knowledge processes in the struggle against declining soil fertility. This case-study is based on the direct experience of Dangbégnon and Brouwers (1990) and Brouwers (1993), who set up the network
- the role of discourse among French dairy farmers as a condition for effective silage making. This case-study is based on research by Darré (1985)
- the introduction of integrated arable farming in the Netherlands. The case-study is based on research by Somers and Roling (1993), and Roling, (1993).

Using this evidence, the chapter formulates a number of working hypotheses with respect to the role of local knowing and its facilitation in sustainable agriculture. It concludes by paying explicit attention to two aspects of facilitation: participatory technology development and the institutional context within which facilitation takes place.

Case studies as evidence

Facilitation of landcare in Australia

Landcare in Australia provides an interesting case of what we are talking about. The landcare movement is a response to the immense land degradation problems which resulted from large-scale use of inappropriate farming methods. In a mere hundred years, European settlers in Australia managed to devastate natural resources on an unprecedented scale. Landcare is based on legislation which creates local landcare groups and provides them with statutory powers. The local groups can thus wield 'the big stick' to bring recalcitrant landowners into line, if necessary. But the basic thrust of the landcare movement is explicitly not to use authority and coercion. The emphasis is on local 'platforms' of stakeholders in a water catchment or other agro-ecosystem, who go through a social process during which they learn to appreciate the land use problems in the area, begin to develop an agro-ecosystems perspective on the natural resources they use in common, and move towards joint action to improve the situation. There are now some 1500 landcare groups in Australia.

It is not our intention here to dwell at length on all the interesting features of landcare. The reader is referred to Campbell (1994). We want briefly to describe a typical instance of landcare facilitation in Western Australia because it demonstrates very clearly the perspective we seek to develop in this chapter: the active creation of local knowledge as a condition for sustainable natural resource management.

In the particular landcare group in question, the facilitator had been appointed by the Department of Agriculture. She was a trained agriculturalist with considerable knowledge of soils, salination and salt-tolerant crops, and wind erosion control. But she had also followed a training course in group facilitation. Without her technical training, the relatively young facilitator would probably have had a hard time with the landowners.

The group came together for the first time and discussed soils. The idea was to develop a soil classification system that was shared by all the landowners in the group. To this end, the farmers went into the field, dug soil pits, made profile peels, and generally agreed on the relevant types of soils in the agro-ecosystem with which they were concerned. The ensuing soil classification system was not the 'scientific' one used in the Department of Agriculture. In fact, the soil scientists in that Department looked at the work of the facilitator with suspicion, claiming that it threatened 'the integrity of science'.

The farmers were given an airphoto mosaic of the whole agro-ecosystem with the boundaries of their properties drawn in. They were also given a transparent overlay and special pens. Their assignment: to draw a soil map of their own property using the soil classification they had developed together. The resulting soil maps were digitized and fed into a computer with GIS software. Thus, the soil maps of the individual properties could be transformed into a soil map for the entire agro-ecosystem. The first time around, the result was bit of a mess. Important soil types would inexplicably stop at property fences! But more discussion and field visits finally led to a product that all land owners agreed upon. The exercise thus achieved an important objective: the landowners had developed a joint perspective on the entire agro-ecosystem. So far, they had minded their individual properties in isolation. Important new shared local knowledge had been created.

This was an necessary condition for the next step: deciding upon a joint management plan. The landowners classified the different soils in terms of their vulnerability to wind erosion and salinity. Vulnerable land management units were identified and management plans made which spanned several properties. This meant that fences, tree belts and so on had to be aligned with a collective management plan: the agro-ecosystem was being managed as an integral whole. Without this, its sustainable management would be impossible. But to reach that point, the active creation of shared local knowledge had been indispensable.

Farmer field schools in Java

After the famines in the 1960s, the Indonesian government launched a Green Revolution programme featuring subsidized packages with high-yielding rice varieties, fertilizers, and pesticides to be applied on a calendar-based routine, all provided on credit. It is important to note that since that time, many local rice varieties, as well as much indigenous knowledge about growing rice and about dealing with pests and diseases, have been lost. Present farmers are the second or third generation users of high input technology, and only those aged well over 50 remember the old ways.

The strategy worked. After being the world's largest importer of rice, Indonesia became self-sufficient in 1983. However, national food security increasingly came under attack by brown plant-hopper outbreaks, resulting from pesticide-induced resistance, and from destruction of the natural enemies of the pest. As a reaction, the government banned 57 broad-spectrum pesticides, declared integrated pest management (IPM) as the national pest control strategy, and abolished the subsidies on pesticides. The IPM programme is based on an ecological perspective and on principles for local application, instead of a Java-wide set of recommendations which farmers must routinely implement. The four IPM principles are: (1) growth of a healthy crop, (2) weekly observation of the field, (3) conservation of natural enemies, and (4) seeing farmers as IPM experts.

IPM required an approach to farmer training that was very different from conventional technology transfer. Farmers could no longer be considered as passive receivers of packages, but had to be seen as active learners and independent managers of ecosystems. After initial efforts to provide IPM training through conventional approaches had proved to be costly failures (e.g., Matteson *et al.*, 1992), a strategy for IPM training was devised which was based on the principles of non-formal education.

Farmers are trained in so-called 'IPM farmer field schools', and it is these field schools that are of interest because of their focus on facilitating local knowing. Groups of roughly 25 farmers with adjacent irrigated fields meet weekly during an entire growing season of 10 weeks. Before the first session, farmers are 'tested' in terms of their knowledge about pests, natural enemies and other relevant knowledge.

Each session starts with a visit to the group's experimental plot, provided by one of the farmers. The farmers are split into groups of five, and each group enters the muddy field at a different place to observe systematically sampled rice hills. At each hill, all insects and larger fauna are observed and, if possible, caught and placed in a small plastic bag. The crop is also carefully observed for stage of growth, health, and damage. After these observations, the groups return to the shed or shade tree and each small group draws the results of the observation on a sheet. The rice plant is drawn, including the salient features of the stage of development, damage, and so on. Pests and natural predators are listed separately, each drawn in carefully and named. Each small group considers the results of the analysis and decides whether pest control measures are necessary. The small groups then present their findings and conclusions to the plenary. In the discussions that follow, pros and cons of using pesticides in the present situation are weighed, information about unfamiliar observations is provided (e.g., egg packets of certain pests) and desirable terms and activities are agreed upon.

Later on, the group engages in a group dynamic exercise and follows one of the prepared training modules, e.g., working out the number of rats one pair can generate in a year by laying out a match for each rat, watching a chicken die of pesticide poisoning, learning about the life cycle of a major pest, or observing the goings-on of pests and predators in the 'insect zoo', a bucket with some rice plants covered by netting, into which various organisms have been introduced. After the last session, farmers are again tested to give them feedback on how much they have learned.

The IPM trainer (very carefully trained for over a year himself, with heavy emphasis on getting into the mud, growing rice and experiencing problems) is considered a facilitator of the experiential learning process, and not an instructor. The training approach enables farmers actively to identify and solve their own problems. A key ingredient in this process is the creation of local shared knowledge: the names of pests and predators, their behaviours and life cycles, the procedures and arguments for monitoring fields and for arriving at decisions about pest control, understanding population dynamics, etc.

Careful evaluation of the IPM farmer field schools (Van de Fliert, 1993) shows that they provided a new perspective on farming for most participants. When asked what they appreciated about IPM, 'having become knowledgeable' was mentioned as often as the financial and other benefits. Farmers gained self-confidence as a result of increased knowledge, which they themselves considered a major reward. In addition, the changed practices of IPM-trained farmers resulted in consistently higher yields in the seasons after training, and in a more favourable yield variability, while money was saved by not purchasing pesticides.

Installing a local network among the Adja in Benin

On the Adja plateau, in Southwest Benin, an effort was made actively to install a network of farmers so as to allow them to learn from each others' experiences (Dangbégnon and Brouwers, 1991; Brouwers, 1993). Objectives of this network were to (cf. Röling and Engel, 1989):

- make valuable farmer practices accessible for other farmers, also outside the area
- identify locally adapted cultivars useful for propagation and selection activities
- identify criteria for technology developement (farmers' objectives and priorities)
- identify detailed indigenous knowledge on local resources, including historical knowledge.

In addition, it was presumed that trying actively to install a network of informants would give information about how farmers communicate among themselves – a valuable basis for anticipating the effects of the interventions of development practitioners to facilitate farmers' learning.

The network was set up as follows. Each participant would receive information from all the others, after providing certain information himself. This information concerned practical knowledge, such as the performance of a new variety, herbs indicating soil fertility status or useful for medical treatment, a new way to store maize, etc. On a monthly basis participants received a leaflet containing the information gathered during the previous month.

Adja farmers regularly undertake experiments. A type of experiment, related to learning by doing, regularly conducted by Adja farmers is called *adokpo* which might be translated as 'trying that, which you have never tried before'. Practising *adokpo* is quite normal among Adja farmers, though with variable intensity across individuals. In principle, an *adokpo* always coincides with a change in the way agriculture is practised. This change might be (1) a personal idea, (2) an idea that has already been tried by another person from the Adja society, or (3) an idea coming from outside the Adja society. Only when a new practice has become routine among peers might it be adopted by a farmer without first making *adokpo*.

Adokpo is mostly applied when testing a new variety or a new technique, such as another way of planting cassava stalks or testing the use of dung. On several occasions during the fieldwork, new varieties of maize, cow pea or cassava were introduced to the villages under study through local processes, and were tested in *adokpo*. This was mostly done by sowing or planting one part of a field with the new variety, which was then compared at harvest time with older varieties in the rest of the field (Dangbégnon and Brouwers, 1990). Another method involved sowing or planting drills of different varieties next to each other. Comparison involves variables like production, ability to suppress weeds, duration of growing cycle, and production of vegetative material ('to feed the soil').

Most *adokpos* are undertaken at the individual level. The results of the *adokpo* remain at the individual level, or sometimes will be shared within a restricted personal network, an *adokpo*-group, with more or less stable patterns of knowledge production and knowledge exchange. The way *adokpo* is carried out gives direct information about the situation and preferences of a farmer.

The effort actively to install a network among *adokpo* groups revealed several learning points. A first conclusion was the great variety of information and knowledge that could be exchanged by actively installing a network. Second, farmers proved to be very heterogeneous. Since knowledge is actively created and pursued only in small *adokpo* groups or at the individual level, exchange is not pursued actively. Farmers do try to observe how others manage their agricultural activities, but this is not done in a structured way. New varieties are often stolen from others, and new practices are learned by 'industrial espionage'. Third, the network should grow only gradually, maintaining active involvement of participants as well as facilitators. Participants gave interesting information only after they had gained confidence in the network during the first stage. Finally, the utilization of the leaflet became practical only after literate youngsters joined in its production and distribution.

In recent years, the Adja have actively adapted their farming system to the changed circumstances brought about by a rapidly growing population and rising demands. Given the near total absence of government activities to develop agriculture in the area, save for the promotion of cotton production to generate export earnings, Adja farmers have done a remarkable job in developing an 'oil palm fallow', an agro-forestry system that uses the multi-purpose oilpalm to restore what the Adja call 'comatose soils'.

Brouwers (1993) describes this system in detail: young oil-palm seedlings are planted in the foodcrop fields. Palms are regularly pruned during some eight years of mixed cropping with annuals – delivering leaves and stalks for burning, thatches, feeding goats, etc. Finally, the field is left to the densely planted palm trees and becomes a 'palmerai' for a number of years. Meanwhile, the fruits are gathered for palm-oil, and leaves for fuel, animal fodder and so on. Finally, the palms are cut down and tapped for palm wine which is distilled into 'Sodabi', a well known and lucrative spirit in the area. (Mr Sodabi fought in the French army during the First World War and learned distilling from his French colleagues). After cutting down the trees, the debris are left in the field and annual crops are planted among them. Thus the cycle starts again. Soil testing by Koudokpon *et al.* (in press) showed that the effectiveness of the system is not based on adding mineral nutrients to the soil, but is better explained by the fact that adding organic material to the soil activates soil life, which in turn releases nutrients from the prevalent clay soils.

The active and rapid indigenous adaptation of agriculture by the Adja is no doubt stimulated by *adokpo*. Experience with the network shows that it is possible actively to enhance such indigenous processes to improve the ability of local people to maintain sustainable farming in conditions of rapid change.

The function of informal discussion among farmers

Darré (1985) studied the effects of informal discussion among French dairy farmers, especially with respect to silage making. What he found was that these informal discussions had important functions that went beyond the diffusion of new information, now often called farmer-to-farmer communication of technology. In the first place, in their informal discussions during birthday parties, along roadsides and in the local café, farmers evaluated technical information coming from the extension technicians in terms of its usefulness from their perspective. In the second place, local discussions created new concepts for dealing with changed perceptions and realities. Darré (ibid.) shows in some detail how concepts, though perhaps having a well-known name, or even when given the name used by the technicians, are invested with new meaning through local informal discussion. In the third place, formal discussion serves to establish the range of acceptable and feasible activities.

In all, Darré's work makes clear that local knowledge processes are constantly at work, even in modern industrial settings, to recreate and re-invent (Rogers, 1995) the technical knowledge being 'transferred' to farmers, and to establish desirable and feasible ways to act. It has been observed that voluntary change presumes wanting it, knowing it and being able to do it: motivation, knowledge and perceived self-efficacy. It seems that informal discussion is an important factor in creating these conditions.

In the French case, the 'technicians' did not deliberately facilitate this local process, even though they played an important role in it. In the following case, local non-formal discussion is deliberately facilitated, although it proves not easy to do so.

Facilitating integrated farming in the Netherlands

Dutch agriculture is said to be incredibly productive. The stamp-size country shares the second place with France, after the US, in terms of the value of its agricultural exports. With only 2 per cent of its agricultural land, the country produces about 15 per cent of the EU's milk, pork and potatoes. But the price paid in terms of surface water atrophiation, ground water pollution, forest die-back, loss of bio-diversity, and destruction of age-old landscapes, eco-systems and habitats has been enormous. Add to this the facts that agricultural markets are saturated, that farm profitability has been negative for several years, and that the continued decline in the numbers of farmers has eroded the political power base of the industry, and one can easily understand the environmental regulation that has recently been approved by the Dutch Parliament. We now have a Nature Policy Plan, Manure Laws, a Multi-Year Crop Protection Plan, with other legislation just around the corner, such as the introduction of compulsory mineral bookkeeping in 1996, and the requirement that glass-houses become closed systems, recycling all their chemical and other wastes.

The regulations only tell farmers what they should *not* do. They do not tell farmers how they can continue to survive as farmers in the new policy context. That problem is left to the industry and its supporting institutions to solve. One of the approaches has been to introduce 'integrated agriculture', which does not completely ban the use of pesticides and chemical fertilizers, but seeks to make their use more efficient (minerals) or minimal (pesticides), while focusing on the utilization of natural processes and on replacing chemical inputs with knowledge-intensive, and to some extent with labour-intensive, practices.

Exploratory research of farmers' reactions and of the introduction of integrated farming (Van der Ley and Proost, 1992; Somers and Röling, 1993; Van Weeperen *et al.*, 1998) has shown that integrated farming represents a totally different technology from conventional farming. Instead of feeding individual crops, farmers must manage nutrient flows across entire multi-year crop rotations, focusing on highly efficient nutrient use which minimizes losses into the environment. Instead of on curative chemical spraying for pest and disease control, farmers must rely on preventive measures based on careful and continuous monitoring and anticipation. Instead of routine and calendar-based application, farmers must use their own judgement and ability to apply general principles in locally specific, if not farm-specific, conditions.

This new technology has very important implications for farming practice.

Farmers must devote a great deal of time and attention to observation and monitoring and must learn systematic methods for carrying out these tasks. They must become adept at using indicators of various conditions and at using new techniques and instruments for making things visible. Typical examples are leaf-stem analysis for fine-tuning the topping up of potatoes with nitrogen, and mineral bookkeeping, a method for measuring all nutrients that enter and leave the farm so as to calculate input/output efficiency (which will later become the basis for a government-imposed levy on inefficiency).

Integrated arable farming has been introduced initially by selecting 35 farmers who were willing to run experimental farms for introducing the integrated farming system developed on a research station. Around each of these 35 farms, 'study clubs' were established of farmers who were willing to discuss the results of the experimental farmers and experiment with introducing integrated methods on their own farms. The 'study club' is a Dutch model for group learning, which has blossomed especially among intensive glass-house horticulturalists who run such complex systems with so many manipulable variables that simple 'extension' has years ago already proved insufficient as a knowledge base for the industry.

The introduction of integrated agriculture has created similar conditions on livestock and arable farms. Instead of professionals and entrepreneurs only, farmers have become managers of complex agro-ecosystems. They must be experts in such management themselves because the integrated sustainable system management required cannot feed only on external experts or information. In fact, one of the fundamental agricultural research institutes in the Netherlands, the Centre for Agro-biological Research, is now developing complex system management knowledge together with farmers, although the mandate of the Centre explicitly prohibits this fundamental research organization from working with farmers directly. It is no longer possible to do fundamental agricultural research without involving farmers' knowledge.

The change from conventional (very) high chemical input farming (Holland is the world's highest user of active pesticide ingredients and nitrogen fertilizer per hectare) to integrated farming is a very drastic and difficult one. We are no longer speaking of the 'adoption of an innovation'. Instead, our studies (e.g., Somers and Röling, 1993) show that a complex learning process is involved. Known risks are replaced by uncertainty. Existing and reliable knowledge and routines must be thrown overboard and replaced by unfamiliar practices. Existing 'theory' becomes redundant and must be replaced by systems thinking and by understanding of complex processes such as the movement of minerals in the soil or the population dynamics of pest organisms under different weather conditions. Decisions must be based on long-term anticipation and can be redressed only with difficulty or at high cost later.

These changes require an immense effort in facilitation of knowledge-building and learning. Extension workers find it very difficult to become facilitators. They are used to being experts who give advice which farmers follow. Now they must be experts, but with a difference. They must be able to stimulate farmer groups to learn together, to develop new concepts, to exchange experiences, and so on – in short, to develop the complex system-management knowledge required for sustainable agriculture. Active facilitation of farmer knowing is a necessary condition for a successful adaptation of Dutch agriculture and for its escape from the dead-end direction of its present development.

Discussion

The experiences described give rise to a number of working hypotheses with respect to the role of local knowledge in the development of sustainable agriculture and its facilitation.

(1) Sustainable agriculture seems to require a 'technology' that is knowledge-intensive and assumes knowledge that was not required before. This knowledge has a locally specific or even farm-specific character. Instead of blanket recommendations which can be routinely applied across a wide area, sustainable farming must operationalize general principles in local conditions. From curative, the emphasis shifts to preventive action. Instead of being routine and calendar based, farming operations shift to being observation and inference based. This requires great attention to making things visible and to conceptual frameworks for interpretation of what is observed. Sustainable agriculture therefore requires the active facilitation of new local knowledge.

(2) Farmer expertise seems an indispensable element in sustainable agriculture. Sustainable agriculture involves the integral management of complex local ecosystems. Farmers' objectives (intentionality), sense making (or reality construction), and self-perceived capacity to make a difference ('agency'), (Giddens, 1987) are essential elements in such management, and cannot be divorced from it. The farmer and his or her local knowledge have become key ingredients in the advance of human knowledge with respect to managing natural resources. Scientific research continues to play a key role, but not any longer as the source of innovation. Research becomes a resource for the farmer to draw upon and to complement his or her systemic knowledge. But farmers cannot effectively rely on external expertise. Sustainable agriculture requires that farmers are themselves experts in managing complex systems. This implies that sustainable agriculture asks for active facilitation of farmer expertise and learning. It also implies that research must actively involve farmers in generating scientific knowledge.

(3) Sustainable management of natural resources often seems to involve creating local knowledge with respect to a higher level of agro-ecosystem aggregation than under conventional management. Erosion control, maintaining biotopes for the natural enemies of pests, integrated rat management (Van de Fliert *et al.*, 1993), synchronization of cultural practices for pest control, the sustainable management of ground water resources, and so on, all require joint management of agro-ecosystems larger than the farm. This holds for industrial agriculture as well as for low external input agriculture. Moving to more sustainable practices thus often means increasing the level of system aggregation of management. This involves the active creation of new local knowledge and information systems required for such management.

(4) Agricultural development can no longer be seen as based on transfer of the products of scientific research and their subsequent spontaneous diffusion among the 'target group'. Change to sustainable agriculture seems to require the capacity to facilitate learning complex system-management. Most of the institutional knowledge systems we have designed to promote agricultural development depart from the assumption that science is the source of innovation. Hence they seek to provide an 'effective institutional calibration of the science–practice continuum' (Lionberger and Chang, 1970). Sustainable agriculture seems to require a totally different knowledge system to support it, with greater emphasis on the facilitation of local learning.

Facilitation of local learning, and its institutional requirements

The working hypotheses generated above draw attention to facilitation of local learning and the design of the institutional environment to support such facilitation. We conclude the article by briefly discussing these two aspects.

Participative technology development and farmer experimentation

There has been a noticable change in the way researchers and extension agents deal with rural people. First, the inclusion of socioeconomic factors was extended to the knowledge construction phase: on-farm technology development. Now an approach is evolving which goes much further and gives a new role to the researcher. Acknowledging the part that rural people play in technology development, the researcher's task is to stimulate ongoing research and diffusion within the rural community itself. This approach is called participatory technology development (PTD). PTD aims to mobilize and ameliorate farmers' capacities to produce and integrate new knowledge and transform this into practical methods and skills, through experimentation and evaluation. Haverkort *et al.*, (1991) described PTD as the practical process of interactively combining the knowledge and research capacities of farming communities with that of the commercial and scientific institutions. Rural people are no longer viewed as passive recipients of external knowledge, but as intentional beings who generate new knowledge and insights, while reacting actively upon environmental limitations and possibilities. A practical methodology for PTD is emerging (Jiggins and De Zeeuw, 1992).

A key ingredient in PTD is farmer experimentation, which may be conceptualized as the linking together by a farmer of cyclical observation, interpretation, and manipulation of his or her environment and the resources at his or her disposal. A farmer is frequently forced to adapt a farming practice because of unpredictable factors. Such 'imposed experimentation' seems typical for farmers obliged to carry out their activities under low levels of control, because of climate variability, social uncertainty, political unrest and so on.

Studying experiments undertaken by rural people allows us to understand their 'sense making' activities. Scientists tend to regard an experiment as an inquiry during which all the parameters are highly controlled except the variables under study. Farmers' 'practice' differs from the scientific experimentation in the sense that it has to be carried out in daily circumstances. Farmer experimentation is an integral part of the whole farming activity. It has to provide direct comparison with adjacent fields and the previous method or technique, while being intrinsically bound up with farmers' conditions in the design, implementation and evaluation of the experiment.

Given its disciplinarian specialization, current agronomic science is not accustomed to dealing with the highly complex, diverse and interrelated knowledge rural people need in order to survive in risky and highly diverse environments. Yet formal science is often seen as the agency that will reveal 'reality' progressively and hence create the technology to deal with it. PTD is based on another view: the social construction of multiple realities takes place via 'sense-making' activities. Knowledge is generated in everyday life via objectification of (inter)subjective processes (Berger and Luckmann, 1967: 20). Rather than seeing human conduct as governed by laws or as caused in the same way as events in nature, we have to understand the intentions and reasons that people have for their activities (Giddens, 1990: 125). Taking this perspective is not easy for the agricultural scientist engaged in PTD.

Cause–effect relationships do feature in the 'environmental' knowledge of rural people (Van der Ploeg, 1991: 212). However, the basic structure of the argument often seems to be different from scientific knowledge: a subjunctive approach which points at a universe of possibilities. 'The subjunctive mood is rooted in the creative fantasy of man and orientated towards the exploration of possibilities and (. . .) selection of preferences considered realizable' (Van der Ploeg, 1991: 212, quoting Van Kessel, 1988). 'Rural people's knowledge is based on different sources of knowledge, such as intuitive preferences, admiration, predilection, faithfulness to tradition, and responsibility.' Accepting this inclusive nature of local environmental knowledge, the existence of multiple realities, the relativated position of scientific knowledge, and the importance of farmers' knowledge in PTD, means a fundamental paradigm change for most agricultural scientists. This means that the shift to sustainable agriculture is likely to have far-reaching consequences, not only for the practice of agricultural research, but also for the paradigm on which the practice is based.

Platforms and agro-ecosystems

The second issue concerning facilitation is the design of the institutional knowledge system (e.g. Röling, 1988; Röling and Engel, 1991). The working hypotheses formulated above suggest that the key ingredient in such a knowledge system is the learning group of farmers, and, at higher levels of ecosystem aggregation than the farm, the platform of stakeholders in a given agro-ecosystem (Röling, 1994). To be effective in terms of sustainable natural resource management, such groups/platforms must have a shared appreciation of the environmental problem situation, must engage in shared learning about the agro-ecosystem in question, develop a common information system at the interface between agro-ecosystem and platform, and develop joint agency with respect to action to improve the problem situation (Checkland, 1981; Checkland and Scholes, 1990). Taking platforms through such a process of social learning requires a decentralized cadre of highly skilled facilitators, both in technical and social terms. Local facilitation must be linked to scientific knowledge and technology development. But rather than a linear 'science–practice continuum', such science linkage can better be conceptualized as conforming to what Kline and Rosenberg (1986) have called the 'Chain-Linked Model of Commercial Innovation', which focuses on a local design process and the creation of access to scientific knowledge throughout that design process.

11. Varietal diversity and farmers' knowledge: the case of the sweet potato in Irian Jaya[1]

JURG SCHNEIDER

Abstract

IT IS ASSUMED that the sweet potato was introduced to the highlands of New Guinea less than 400 years ago, yet today an amazing number of land-races, or local cultivars, are found in the area. Sweet potatoes are the staple crop in both Irian Jaya, Indonesia's eastern-most province, and Papua New Guinea (PNG). Research to collect, study and preserve these land-races in Irian Jaya is underway, but the concepts that we use to understand how varietal diversity is created, maintained and transformed by farmers are still inadequate. This paper looks at some scholarly hypotheses that have tried to explain varietal diversity of sweet potatoes in Irian from either a sociological or a biological perspective, and at the shortcomings of these disciplinary approaches. It is concluded that a better understanding of the traditional, autonomous mode of variety selection will be important not only scientifically, but will also help to support the conservation of biological diversity of food crops *in situ*.

Introduction

Early evidence for agriculture in the New Guinea highlands dates back to at least 3000 years BC. Developed agricultural systems based on taro and a number of other crops existed long before the diffusion of sweet potatoes, which started about 400 years ago. Botanical and genetic studies have shown that sweet potato (*Ipomoea batatas*) is of American origin, with a probable centre of origin in western South America (Yen, 1974: 245). The view most commonly held is that *Ipomoea* arrived with the Iberian exploration of the Indonesian archipelago (Yen, 1974: 317) and diffused from the coastal part of New Guinea to the interior, where it arrived considerably later (e.g., 250 AD in the Wahgi Valley, PNG) (Golson and Gardner, 1990: 407).

Given the short time of its presence, the current dominance of sweet potato in almost all parts of the highlands is surprising. The major cause of the rapid adoption and spread of sweet potato has been seen in two factors where it outcompetes particularly taro: first, its higher adaptability to environmental risks such as drought and low temperature, and second its higher productivity combined with its great suitability as a pig feed both in cooked and uncooked form (whereas uncooked taro is not eaten by pigs). The sweet potato enabled highland societies to expand settlements into higher altitudes (much above 2000 m) and to shift their subsistence base from hunting more towards horticulture, and to raise more pigs.

---

[1] This paper draws on research carried out in co-operation with the Central Research Institute for Food Crops (CRIFC), Bogor, and the Root and Tuber Crop Research Center (RTCRC) at Universitas Cendrawasih, Manokwari, Irian Jaya. I would like to thank especially Dr. A. Dimyati, Co-ordinator of the Indonesian National Root Crop Research Program, Ir. Hubertus Matanubun, Chairperson of RTCRC, and C. A. Widyastuti, Research Assistant (CIP), G. Prain (UPWARD), and Ir. LaAchmady, Jayapura.

The importance and diversified use of sweet potato is reflected in varietal diversity. The number of local cultivars may serve as a first though rather crude measure of varietal diversity. By local cultivar (also often called 'land-race') we mean a cultivar distinguished, selected and named locally. In Papua New Guinea alone, about 1000 local cultivars were collected and are maintained on-station, and many more are known to exist. In our own collections (Schneider *et al.*, 1993), we also found that the number of cultivars is impressive. In the Baliem valley in the Central Highlands (Jayawijaya regency) with an area of less than 1000 sq km, at least 200 local cultivars occur, and about the same number of cultivars is found in the mountainous area to the west of the great valley. An exploration in the division of Paniai (western highlands) collected 139 cultivars (Matanubun *et al.*, 1991).

Explanations for a problem

Name counts tell us something about diversity, but what kind of diversity? In other words: Why are there so many names?

Karl Heider, an anthropologist who had done research among the Dani in the highlands of Irian Jaya, was the first to address this problem. Clearly, he was wondering about 'the proliferation of names' and found it unlikely 'that any one person could manage with any consistency a set of 30 or 50 or more complementary categories in a realm like sweet potatoes' (1969: 79). He went on to suggest two potential reasons for the long cultivar inventory, but admittedly failed to give conclusive evidence for any one of them.

- Names proliferate for cultural reasons; there are a great number of microdialects, for example, in the Dani language area. Each community may choose different names for 'their varieties' which of course increases the total number of names without increasing the number of morphological types. Heider also suggested that sweet potato varieties represent an opportunity for playful naming so that one community may have several names for one single variety.
- Varieties are named because of functional differences, their different performance and distinctive adaptations to a number of micro-environments or natural stress factors, such as flooding. However, he found it impossible to prove it with the scarce information extracted from (male) farmers – information or knowledge that would relate specific varieties to specific 'functions' or constraints.

Heider's approach has a social science bias, in that it remains in the sphere of names as cultural symbols, and makes no attempt to look at plant variation as the natural base of perception. Plant geneticists have developed detailed lists of descriptors (such as 'leaf form') for many crops, including sweet potato, which allow us to characterize varieties, and to distinguish very minor differences. Our own characterization work (on the experimental station where the varieties collected are maintained) has shown that the great majority of names refer to distinctive varieties. This result tells us two things: first that there is not as much 'playful naming' as Heider seems to have assumed, and second, that local farmers have a very sharp eye for distinctions.

Very much aware of the genetic dimension of the problem of varietal diversity is Douglas Yen (1974) who did an exhaustive study of the sweet potato in Asia and the Pacific, and devotes several pages to sweet potato in Irian Jaya.

A big inventory of local names does not imply that the respective cultivars have a broad genetic base. In the case of Irian Jaya, which is a secondary centre of diversity, the genetic base of sweet potato is most likely smaller than in South America. What we can say, however, is that the number of varieties originally introduced must be far below the current number of landraces. Although all these landraces might have a common gene base which is smaller than that of South American sweet potato, their differentiation is an achievement of local plant selection. And, surprisingly, this aspect concerning human action in and impact on varietal diversity is completely missed out by Heider, but receives a certain amount of attention from Yen. How does this plant selection work?

- Plants are reproduced vegetatively (through replanting of cuttings) which means that the genetic identity of a cultivar or genotype is preserved. The system of propagation common all over the highlands does not support genetic recombination. Yet existing varietal diversity is evidence of such a process.
- Sexual propagation through seed is incidental, but it is favoured by continuous cultivation which always has some old gardens where the plants produce seeds. Flowering and seed-setting is frequent, and these seeds are new genotypes. Farmers do not exploit such seed systematically to get new cultivars. Yen thinks that they just don't know about sexual reproduction of sweet potato (1974: 231). This may or may not be true, but farmers are certainly aware of one way by which diversity develops; through volunteer seedlings. In older gardens, seed is likely to germinate and to produce seedlings. Those seedlings are a field of experimentation, and may in the end become new varieties.

Among the Eng in PNG, the evolution of new varieties is attributed to certain bird species (Bulmer, 1965). Some Dai variety names reflect the appearance of novel varieties, and probably the farmers' awareness of the changing nature of their planting material. There is a name indicating the time when a variety was encountered during garden work (*linggoara*, 'found at noon'), or another referring to the bird that dropped it and thus may be considered its procreator (*tuwenekara*, 'dropped by the bird'). That varieties evolve in the process of cultivation is reflected by the subdivision of variety names into subvarieties called the 'new' and the 'old' cultivar x (Schneider *et al.*, 1993: 38).

Towards integration of social and biological approaches

For a better understanding of varietal diversity, the genetic and the social dynamics need to be analysed together, with a grasp as good as possible of the part of the indigenous knowledge relevant to it.

We should also look not just at varieties, but also at populations or cultivar mixtures. Once we realize that every household manages a mixture of a dozen or many more cultivars in several gardens of different age, this becomes an important issue, because planting material (vines) flows between the garden sites. It is a matter of how much of which variety. Intentional selection for distinctive properties takes place, but also unintentional selection because the favoured variety is not available. This is an indigenous (and still autonomous) mode of variety selection. It is indigenous because it is mostly driven by subsistence needs, and to a very little extent by market forces, and it is autonomous because varieties are of local stock as opposed to variety introductions from institutional agricultural research. Yen (and others, including myself) have as yet very little data on this aspect to contribute.

To understand this autonomous mode of variety selection better and to make it bear on agricultural development, we have to combine, as best we can, three types of information.

- The local phenotypic and genotypic variation of the crop.
- The indigenous knowledge directly related to cultivars.
- The social aspects of garden management and variety selection.

In a study on a manioc-growing community in the Amazon, Boster (1985) has combined the first two aspects very well. He came up with some fascinating interpretations. His findings indicate that the number of basic uses is much smaller than the number of varieties. The 60-odd varieties correspond to only two basic uses: manioc for beer and manioc as a staple food. The beer-making varieties grow more rapidly, produce larger roots, have more fibre, and rot more rapidly after harvest. The other group contains the varieties boiled for eating.

Yet, the actual number of varieties is much larger, each cultivar being mainly distinguished on the basis of leaf morphology and stem. These are most often gradual differences, not marked ones, such that the outsider even has difficulties in 'seeing' them. Why are so many maintained? A (traditional) breeder knows the answer of course. Keep anything that looks different, and for an Aguaruna, it might not be too different.

If we are trying to understand, we thus should not take an entirely utilitarian approach. Many characteristics are useful for making distinctions, but do not automatically mark an important functional difference. Selection is working first on the level of perception, and only second on the level of utilization. Boster called this 'selection for perceptual differences' (SPD). SPD provides the farmer with the raw material to work with.

In a subsistence production system, such as the one prevailing in Irian sweet potato farming, these are important considerations. There are multiple uses of the sweet potato, and the goal is not to select the best one, but to get an appropriate, well-performing mixture of cultivars from the three groups I mentioned at the beginning.

Development and conservation of a local resource

We have taken only a glimpse at the complexities of a small part of an indigenous knowledge system – the ethnobotanical knowledge of highland horticultural societies in Irian Jaya. I would like to make two remarks on how this is related to the overall theme of this book, and what is, I think, the rationale of a better understanding of this knowledge.

What we have seen suggests very strongly that Irian farmers have adopted sweet potato in historically recent times and used its genetic potential to adapt it to a varying mountain environment. This knowledge is valuable and relevant because it allows farmers to distinguish varieties and to select for distinct properties. What is 'development' going to do to this skill? At the present time, sweet potato is not paid much attention in the agenda of agricultural development in the region. Varietal change and introduction are almost totally determined locally. This is very fortunate one might say because farmers, not researchers, are directing variety selection, and they have not been separated from the full knowledge associated with a particular plant genetic resource.

However, farmers in Irian as everywhere else have to adapt their cultivar inventory constantly, for example to new pests, and they have a keen interest in testing superior varieties. It is likely that they readily accept varieties from formal plant research and breeding. At present, their gene pool is, we assume, small compared with the gene pool available to plant breeders. The success of 'improved varieties' has depended mainly on two factors: first the control of the environment in which it is planted, and second the availability of a larger and well-characterized gene pool.

The weakness of improved varieties is often the lack of specific local adaptation. This is the strength of a system like the various Irianese horticultural systems and, as we see it, an area of co-operation between formal and informal knowledge systems which can be developed.

One model to put this into practice could be named 'curatorship plus participation in variety evaluation'. Curatorship basically refers to *in situ* conservation, which has become a buzzword recently, but as often as not it is unlikely that farmers will be interested at all in this new method prescribed for them. Obviously, they will not be interested in conservation *per se* unless their personality is that of a conservationist. I see some chance, however, in combining modest conservation goals with farmer-driven evaluation of both their traditional and improved material. This is a participatory research activity, which has been tried out in international agricultural research.

In the case discussed here, farmers would not only be involved at an early stage in variety evaluation, but also experiment with 'varieties in the making', or populations that contain promising traits. The major difficulty of such an approach will be to strike the balance between the old and the new. To those who think that is a Utopian thought, a potato breeder once assured me that many successful potato varieties have been evaluated in this way. But of course, it is a Utopian thought for another reason: I can give you no example for the crop and the area I have been describing.

12. The indigenous concept of experimentation among Malian farmers

ARTHUR STOLZENBACH

Abstract

IN SANANDO, MALI farmers have a concept of *shifleli* that is related to experimentation. The nature of *shifleli* is presented making use of some clear cases. But it appears not to be so simple and clear-cut, when *shifleli* and experimenting is seen in the broader context of innovation. Different dimensions to experimenting are problematized and the Western concept of experimentation versus *shifleli* is relativized. This brings up issues about constraints and opportunities for intervention in indigenous experimentation.

Introduction

This chapter tries to achieve a better understanding of farmers' experimentation and brings up issues on constraints and opportunities for intervention. It is an elaboration of a contribution to the IIED/IDS seminar 'Beyond Farmer First' (Stolzenbach, 1992) and an article in the ILEIA Newsletter (Stolzenbach, 1993). There is a need for such an understanding. 'Farmer-First' (FF) seeks to strengthen the experimental and innovative capability of peasant farmers, in order to enhance the generation of appropriate technology; innovative activities involve transformation of knowledge and technology. When the user develops his own technology the transformation of technology is reciprocal to the transformation of his knowledge. A fundamental weakness of the TOT (transfer of technology) paradigm is that this reciprocity at the level of the user is being detached when the technology is developed apart from the user. Thus, FF seems correct in stressing the importance of the innovative capacities of peasant farmers for appropriate technology, but it is not proved or theorized 'why FF interventions should be any less likely to disrupt or undermine indigenous patterns of technology development than conventional interventions' (Gubbels, 1992: 41).

To promote FF as an alternative, or at least complementary to, the paradigm of TOT, the literature on FF has been primarily focused on advocating what *should be done*. The weak point of FF is that there is much less effort in providing a supporting theory of explanation, that analyses what *is being done* within a conceptual framework that helps to identify the necessary and sufficient conditions for the emergence of the FF paradigm (Gubbels, 1992: 34).

> It is only on the basis of such an improved understanding of farmers' logic and methods of experimentation, and enhanced local recognition of the value of local experimentation, that an outsider effectively can assist in finding ways to overcome limitations in the actual experimental practices and the organization of experimentation and sharing of results within and amongst the communities. If not, one might easily fall again in the trap of the outsider imposing 'improvements' (in this case in farmers' way of experimentation) that are not sustainable within the local socio-cultural context and that will be left aside again as soon as the outsider withdraws (ETC, 1991).

The empirical material is based on fieldwork done in 1991 in Sanando, a semi-arid region of Mali. The agriculture in this area is typically complex, diverse and risk-prone. The research took place in villages where World Neighbors (WN) was active. WN is a grass-roots organization with a participatory technology development approach, that among other things introduces simple innovations, stimulates and assists farmers to experiment, and organizes meetings of farmers from different villages (Gubbels, 1988). They asked me to investigate the nature and sorts of experiments undertaken by these farmers themselves.

Since the aim of the research has been to explore new concepts and eventually build new hypotheses, qualitative methods were applied in the field – for the most part, unstructured interviewing. The starting point was that the farmers seemed to have a word that relates to experimenting: *shifleli*. So I started interviewing using two strategies that also made my own bias more detectable. In the first place I asked if they had done *shifleli* and, if so, we discussed the 'hows', 'whys', 'whens', etc. On the other hand I tried to reconstruct how they got the knowledge they had and the processes of changing the techniques applied. The motivation for this approach was that experimenting, in no matter what form, can lead to changes in the techniques applied. By identifying these changes and reconstructing how these changes had taken place, it could indirectly be distilled which experimenting activities they undertook. At the same time I did not have to introduce my own concepts of experiments and experimentation to them, which, although only vaguely aware of, I did, of course, have. Trained as an agronomist my view was something like: an action, undertaken to learn explicitly from it, and consequently undertaken in a particular way to be able to learn most from it.

Soon I ended up with mostly elderly men, because they are the ones who co-ordinate the farm and cherish the agricultural knowledge, a valuable property, that is not easily shared with the young men.

The nature and logic of *shifleli*

When farmers were asked what *shifleli* meant to them, typical ever-recurring elements mentioned were: close observation, show or prove something to others and check what others say. Also mentioned was comparison of something known to something unknown.

The range of themes is broad, mostly based on an appreciation of changing situations and opportunities. One test, for instance, concerned the proper sowing date of an unknown variety of cowpea (*Vigna unguiculata*), because cowpea is very susceptible to drought or excess of rain, especially at flowering. Another test was to see if the harvest would be better preserved in the granary after treatment with a certain insecticide. Because crop residues of plants such as straw are scarce nowadays, the granaries are built from loam. This results in greater post-harvest losses caused by insects. But most of all, *shifleli* concerns the test of new varieties.

When asked for criteria of *shifleli*, farmers were quite vague, not specifying much more than that it must work in real-life situations. However, in the stories of how cases of *shifleli* had evolved, farmers put forward implicit criteria. And it appeared that these criteria can be linked to different kinds of experimentation, as distinguished by Schön (1983) in his study of the rationality of practitioners. Schön distinguishes three kinds of experimentation, each with its own logic and criteria for success and failure.

- When action is undertaken just to see what will be the results, without predictions about the results, he speaks of an *exploratory experiment*.
- In the *hypothesis-testing experiment* there are already expectations about the results of the action. The purpose of the action is not to change the environment itself, but to test the assumptions underlying the expectations. The experimenting succeeds when competitive hypotheses that try to explain the same phenomenon are proved inferior.
- The purpose of the action in a *move-testing experiment* is a certain desired change in the environment itself. As an example Schön mentions the move of a chess-player, who moves his pawn (only) with the purpose of protecting his king, although he cannot foresee all the consequences of this move. The experiment is successful if the results of the action, with all its consequences, are considered positive, although the underlying hypothesis and assumptions may be incorrect and they might not have been expected beforehand. Then the move is affirmed – if not, then the move will be negated. If, for instance, our chess-player accidentally mates his opponent by this move, he would not withdraw the pawn because the result is not as he had expected.

Let us consider two cases of *shifleli*, the second of which is given in Box 1. Solo Keta had sown two plots with groundnut. The plots differed only in the application of fertilizer: one plot had not received any manure at all, the other had received mineral fertilizer. In the fertilized plot the vegetative growth of the groundnut was very much stimulated, as he had expected from what he had seen before with cereals. But in this particular case he became anxious that, after flowering, the gynophore (the downward elongating peg that contains the growing seed) could not reach the soil and thus would not produce seeds. He intervened by earthing up the plants of the fertilized plot.

After the harvest Solo was very satisfied with the yield increase on the fertilized field. However, the bad taste did not please him. This would not be very problematic if he were to sell it, but for him the market for cotton was more interesting than that for groundnuts. In the end, he decided not to continue with the application of fertilizer, because it was not worth the cost of the fertilizer and the extra labour of earthing up.

When Solo Keta starts his groundnut experiment he takes the first step to gradually change the management of his farm. Very soon he has to reconsider the effects of his action and intervene during the process in order to achieve his vaguely defined problem-statement 'a more successful groundnut production with the application of fertilizer'. This move-testing experiment is completed when he negates the result in all its consequences and decides not to continue with this idea. In Box 1 a farmer from Koyan does his move-testing experiment in order to be able to harvest earlier in response to the climatic changes. After having harvested, prepared and eaten the new millet variety he affirms his move. Now he changes the problem-statement by decreasing the distance between the plants and continues his move-testing experimenting.

The moves of Solo and the farmer from Koyan can also be explained as an exploratory experiment. Their moves 'stimulate the situation's back-talk, which causes them to appreciate things in the situation that go beyond their initial perceptions of the problem' (Schön, 1983: 148). Solo had not realized that the ovaries might not be able to reach the soil. To the farmer from Koyan it appeared worth trying to decrease the planting distance of this variety.

In their statement of the problem Solo and the farmer from Koyan make a lot

Box 1: *Shifleli* on a new variety

The first time a farmer of Koyan had seen '*sunan*' (a small variety of millet) he approached the owner and was told that this small variety of millet can be harvested early and yields well. Since the length of the rainy season had been decreasing over the last few years, he was very interested and he received a handful of seed to try out.

Back home he decided to sow at the shortest distance the people of his village used when sowing millet, so he sowed at a distance of four hand-widths. The new variety did produce well, as the other farmer had said, although 'the taste is not so good and the colour when it is prepared is a little bit black'.

Probably the yield could increase by decreasing the plant-density and so he reduced the sowing distance the next year. This time he was sowing it on large plots and each year he reduced the distance a little bit, until one year the distance had become too short. At the end, the optimum on his fields proved to be more or less two hand-widths.

of assumptions and an (implicit) hypothesis. The results of the experiment can confirm or disconfirm these and in this way it also becomes a hypothesis-testing experiment. Solo's hypothesis that fertilization of groundnuts can increase the yield is confirmed. The assumption that fertilized groundnuts can be cultivated in the same way as non-fertilized groundnuts is disconfirmed. In the case of *sunan* the assumption that the sowing-distance of four hands gives best results is disconfirmed and leads to a new hypothesis that closer is probably better.

Criteria for success of experiments

By earthing up, Solo Keta confirms his hypothesis that fertilized groundnuts do yield more. This contrasts radically with the scientific way of hypothesis-testing through trying to disconfirm the hypothesis. When we consider the very different praxis of the farmer and that of the scientist, this can be better understood. The farmer himself is part of the situation under study and it is he himself who has a direct interest in improving this situation according to his wishes. Agronomists have to describe their findings, and methods should conform to formal systems of explanation (although in practice they often do not). In the scientific model of problem-solving, learning and decision-making are separated because this method requires that formal, rigid analysis be used to gain insight in the situation, so that the best alternative can be executed. To the farmer this analysis is often far too rigid, and besides it can begin only when the complex reality is reduced to clearly defined problems, a reduction that in the farmer's practice often will lose its relevance rapidly. To him it is not relevant nor achievable to explain how the production changes as result of the separate factors. The farmer's interest is to understand the production process in its real complex production circumstances.

The farmer understands the situation by trying to change it and reflecting on the results. His reflection is on the one hand subjective because it is deeply rooted in his appreciative system; on the other hand it is also objective because it can be tested continuously on the basis of the phenomena. Of course, the farmer considers it worthwhile at that moment. So, although a farmer attempts to realize his wishes, deliberately trying to prove his assumptions, this still does not mean that he is creating self-fulfilling prophecies, because the environment resists total manipulation, and gives feedback. But it must be acknowledged that some farmers are more receptive than others to feedback.

Coming back to *shifleli* we see in general that when farmers have tested the technique on a small field for the first time, this usually gives them enough information to reject the technique or try it out the next year in a larger field, under slightly different circumstances. In the next section it will be seen that farmers will look for explanations when the new technique does not work out satisfactorily. Factors that unintentionally positively influence the result are not regarded as a disturbing interference. Whether they can find an explanation can also influence their further actions, for instance in the case of Salia Diarra given in Box 2.

Box 2: Repetition of an experiment

Salia Diarra is one of the few farmers who repeat an experiment on a small scale. Like the others, he will sow a new variety that has performed well in his first small-scale test, but the next time it will be in a large field, irrespective of the experimental conditions. But if the new variety inexplicably does not perform well when Salia thought if would, he rejects the variety and will not consider it further. If the performance of the variety is bad, but Salia can attribute this to unfavourable conditions, he will repeat the experiment another time. Schematically his evaluation is as follows:

Judgement of the result of the experiment	Explicable with local knowledge	Action on basis of the experiment
desired	yes	variety accepted
desired	no	variety accepted
undesired	yes	variety rejected
undesired	no	repeat test

This reveals that the hypothesis-testing aspect of experimenting is less important to farmers than the move-testing aspect. This can ensure that, in contrast to the scientific method of experimentation, the changes in the run of the experiment are not faults, but the essence of the success of the experiment (Schön, 1983)!

Explanation of effects

One of the difficulties of agricultural field experiments is that the observed phenomena are not only determined by the objects but also, and to a large degree, by a multiplicity of circumstances that one cannot control (Hoveyn, 1991). In the case of Sanando, the inconstancy of the rain, both in quantity and timing, variation of soil fertility, erosion and losses caused by birds and insects, will very much influence the results of agriculture.

Agronomists try to separate effects using statistical analyses and they design and execute experiments according to the requirements of such an analysis. The main point of this design is 'to limit oneself to the framework of the research problem by means of a thought-out choice of factors, etc.' (Hoveyn, 1991). If, in the course of the experiment the framing of the problem and thus the execution of the experiment changes, the analysis is no longer valid.

Although farmers do change the execution of their experiment during the course of it, most farmers claim, if the production was disappointing, to be able to determine the limiting factors. Of course, their standards of accuracy are lower than those of agronomists. Actually, if differences in yield are clear to the eye, no need for measurement is felt. Besides, most farmers cannot calculate, and even if they could it would be difficult to convert yields to yield per acre, since the area of the fields are often not known. Some farmers, like Solo Keta, who have contact with extension agents, always do measure the yield of their experiments, but in general the destination of the yield is most determinative, as the next case in Box 3 will show.

Box 3: Selective weighing of the yield

In an experiment with the new varieties of cowpea, a farmer weighed the yield of all the varieties the first year. He also had them prepared and tasted them. He decided to sell the variety with the highest yield but with a bad taste, and to consume the variety that had a good taste and an average yield. The next year he repeated the experiment with a lower plant density and weighed only the yield of the variety destined for sale.

So the farmers' quantitative analyses is quite rough. But the strong point of the farmers' perception is the frequent observation of their crops during the whole season. Parameters often mentioned by farmers are: vigour and colour of the stem, colour of the leaves, length of the plant, size of seeds, time of germination, ripening, etc. Retrospectively they can determinate factors that could have influenced the yield. For instance, it is a good beginning to the growth when the lower parts of the stems are dark green or grayish. When it is reddish there is probably a lack of water and/or low soil fertility. The environment is also taken into consideration. The kinds of herbs, and their growth and regrowth, can be related to soil fertility. On the other hand the weaker part of farmers' perception is that it is limited to directly observable phenomena. So the buds on the roots of beans are only considered as 'a special way of growth'. Nor is soil fertility further specified by categories, as is for example the Western system of analysing minerals in the soil.

Comparison of different locations is also an essential element in determining causes. Not only *between* treatments, but also within a treatment. For instance, if the average yield of a plot is not satisfactory, but there are spots where the plants do grow and yield well, it can be concluded that it is not the rain but the soil fertility that has been the most limiting factor. If the production in the whole field is low, probably the rain has been the limiting factor. To farmers, spontaneous variation is a source of interpretation.

Comparison to a reference makes interpretation easier. Box 4 gives an example of how keen perception and deduction can create standards for comparison.

Box 4: A certain variety as reference crop

Lassana has one variety of beans that is very sensitive to rain. When conditions are favourable it produces more than his other varieties. But if, during flowering, there is a cloudburst, the flowers will drop and the crop will not form. In the case of shortage of rain, the sunlight will wither the flowers and it will not produce either. Lassana uses this characteristic: 'Because this is the most delicate variety, I can know why another variety has not yielded by referring to this one'.

Experimenting as performance

These cases of *shifleli* could be easily identified as such, because they were somehow isolated from the principal production in place and/or time. Then, for a time, it seemed that I saw more *shifleli* than did the farmers, for instance, in Adama Diarra's yard.

Box 5: 'Just' mixing varieties

In a corner of his yard, Adama had sown beans of a new variety. On the other side he had sown last year's beans at double spacing between rows. One month later, in between these rows, he had sown another of his varieties of beans. He told me that this year he did *shifleli* in the corner of the yard. But although he had never at the same time mixed two varieties of beans and sown them in between each other he did not consider that to be *shifleli*, because 'he already knew the varieties of last year'. This year 'he just tried to spread the time of the harvests'. Accidentally he had had two varieties at his disposal and found it 'interesting to mix them'. After discussion he agreed with me that 'indeed you can call it *shifleli* if you want to'.

Farmers do not classify this last case as *shifleli*, because it is completely integrated in the production process and more driven by intuition than by an explicit desire to learn. The move-testing aspect of experimentation is dominant. Nevertheless, to me, it comes close to an experiment, although it may be more similar to 'just' experience. In this case the criterion of purposeful action for learning is problematic, especially because Adama has different purposes at the same time for the same actions.

Where does an experiment start and where does it end? Maybe it never ends, and it is arbitrary to set a limit. Experimentation is inherent in agriculture, because to practise agriculture means doing, judging and adjusting.

Seen in this way, a farmer is improvising on a repertoire of different intertwining themes. Richards (1987) used the term 'adaptive performance' for it. The way this performance improves is learning. As such, experimenting as a continuous innovative element of the craftsmanship of farming is a way to learn in practice. The detailed agricultural knowledge of other farmers who told me they never do *shifleli* made me realize that on the farms in Sanando there are so many spontaneous situations in which one can learn by discussion or by mere observation that the importance of experimenting for learning cannot be overrated. For example, it often happens that two different farmers on adjacent fields are cultivating the same crop, each one in his own manner. So, when harvesting, which they will often do together, each will know the other's results and can learn from them. Also, different people working on the same field can apply different 'treatments'. For instance, children may sow at shorter distance because they have short legs, or 'because they have not understood the instructions properly'.

Box 6: Agriculture as a performance

As soon as the first rains start, the soil is being tilled and when the rains continue the seeding starts. If it stops raining a few days after germination it can be necessary to sow the fields again. Maybe the farmer will now choose another variety or maybe he does not have any seed left.

Experimentation as a learning setting

Since the nature of farming is adaptive performance, farmers' experimentation, even when explicit, is not very systematic in general. The optimism of the FF approach to 'improve' farmers' experimenting or to do experiments that serve the individual farmer as well as the agronomist seems, in the context of Sanando, overdrawn. It is crucial to realize that farmers have to deal with a very complex practice-context in which the scientific way of research is limited in its problem-solving capacity, even when it is done on the farm. The goals of farmers and agronomists are different. Scientification of farmers' research, would miss the point (Van der Ploeg and Douwe, 1987). Flexibility and adaptive performance, essential qualities for the farmers, do not easily go together with systematization.

But apart from the development and exploration of techniques, experimenting can also serve as a linkage-mechanism to facilitate communication. In demonstrating and discussing experiments, participants are stimulated to make (the implications of) their knowledge explicit and thus exchangeable. There are barriers to exchanging knowledge and as a result the informal research of farmers is mostly limited to the unit of production. Knowledge is exchanged mainly within the extended family. The experience of WN in Mali shows that development organizations can have an intermediate function and bring farmers into contact with each other. In fact, WN actually linked their intervention to the already known concept of *shifleli*. Within the setting of a project of social learning some barriers can be taken away (like the fear of theft or witchcraft). However, the question remains how sustainable these networks are.

Conclusions

Without having full insight into the farming situation, farmers on the one hand have to take decisions and actions to achieve their goals, and on the other hand to reflect on the results in order to be able to improve their performance. This adaptive performance is primarily based on reflection upon chains of related operations *during* the production process. Seen from this point of view the management of a farm can be regarded as a continuous series of experiments, by which, through the labour itself, the performance improves. A crucial element in this is the ability to reframe the problem according to the changing situation and act according to it, instead of following a 'thought-out design', which is more suited to testing a hypothesis thoroughly in the way agronomists do.

This implies that experimenting must be considered as a *continuous innovative element* of farmers' craftsmanship. It can be isolated from the production process, both in time and space. Then it is given a name, like '*shifleli*'. But it can also take place completely integrated in it. Then it is most often regarded as 'just' experience.

Experimental activities of farmers can have, by the same actions and at the same time, a explorative, an hypothesis-testing and a move-testing function. Since agricultural production is more important than research to the farmer, the rigidity of the hypothesis-testing is less important than the move-testing. This prevalence of the evaluation of the move against the farmers' norms and interests, means that farmers' experiments are very subjective. On the other hand, the possibility of testing the hypothesis implies a certain objectivity in the experiments.

Unlike scientists, farmers do not pin down the design and execution of their experiments but adjust them during the run of the cropping-period. These changes are not considered faults of the experiment, as they would be in a formal scientific experiment. Because of the preponderance of move-testing to hypothesis-testing, but also because spontaneous variation is considered a valid source of information itself, it can even be the essence of the success of the experiment.

The importance to farmers of experimenting as distinct from production should not be overrated. On the farms in Sanando there are so many spontaneous situations from which one can learn that the role of experiments in learning is limited.

Farmers' experimentation has its own strength. Improvements using Western methods must be looked at critically, since the differences extend further than differences in methods. From the view of the outsider, experimenting can also have a function as a linkage-mechanism. Since experimenting is typically learning by doing, and can also be a concrete learning setting for discussion, it is an interesting instrument for *social learning* and demonstration, that can remove some barriers to exchange of knowledge.

13. *Umnotho Wethu Amadobo*[1]: The clash between indigenous agricultural knowledge and a Western conservation ethic in Maputaland, South Africa

DAN TAYLOR

Abstract

THE INHABITANTS OF Maputaland have survived for centuries by utilizing natural resources in a sustainable manner. The fact that the area retains its natural beauty and biological diversity bears testament to this statement. This has led to conflicting land-use alternatives, and a clash between traditional farmers and conservationists, both of whose perspectives have strengths and weaknesses.

Farmers use a variety of methods and habitats to ensure a harvest in an area which is marginal for agriculture. In addition to agriculture, survival has been ensured through pastoralism, hunting, gathering, fishing and more recently migrant labour.

Modernization has challenged a way of life which has ensured low levels of malnutrition relative to elsewhere in KwaZulu[2]. Maputaland has experienced the effects of numerous development initiatives ranging from large-scale agricultural schemes and forestry plantations, to small-scale agricultural projects. Such schemes continue to be presented by development planners and include state conservation initiatives which may involve private sector participation. Two cardinal features characterize most of these initiatives; namely a failure to incorporate indigenous knowledge and the enclosure of the commons.

This chapter will describe the agricultural practices of farmers in one area of Maputaland and demonstrate the dynamic nature of these agricultural systems. It will examine the consequences of a conservationist intervention which pays inadequate attention to the needs of farmers, and suggest a strategy for resolving the problem. Solutions will build upon indigenous knowledge and agricultural systems practised for centuries.

History of the area

Agriculture is not simply a product of biophysical elements but of the political and socio-economic forces that interact with farming systems, together with the history and culture of the farmers themselves.

Maputaland (also known as Thongaland) stretched from Lake St Lucia in the south to Delagoa Bay in the north. Archaeological evidence reveals continuous occupation since the thirteenth century by the Thonga people whose identity as a tribe or ethnic group remains unclear (Webster, 1991). Thongaland has never been ethnically homogenous. Felgate (1982) refers to the people as Tembe-Thonga – derived from the Tembe who dominate in north-eastern Maputaland and the Thonga (Junod, 1927), the majority of whom reside in Mozambique.

Thonga power reached a peak in the mid-eighteenth century when under Mabudu of the Tembe clan, they were strong militarily and prosperous as traders. By the end of the century they had been eclipsed by both the Zulu and Swazi peoples both of whom had been constituted into 'nations'. The

Thonga were never conquered militarily, since their low-lying, swampy territory, ridden with both malaria and nagana (tsetse fly) was unappealing to the Zulu, who were content to exact tribute.

With the decline of the Zulu empire in 1879, Zulu dominance came to an end. Thongaland was divided by the colonial powers (without any reference to the local inhabitants) under the MacMahon Award of 1875 into a northern section under the Portuguese, and a southern section under the British. This had little initial impact on the local people. Their allegiance to the Thonga chief who resided across a colonial boundary to the north, continued. In fact it was only in 1887 when the Queen regent Zambili sent a deputation to the British to complain about Portuguese encroachments on their land that they were informed about the decision (*Zululand Lands Delimitation*, 1905: 288). In 1896 Ngwanase, the Thonga chief, fled across the border after a dispute with the Portuguese authorities, and established control over the southern portion. The area subsequently was administered as part of Zululand, a situation that remains to this day.

As Webster (1991: 249) says, 'there was an autochthonous population augmented by immigrants over a lengthy period. This population probably had little political cohesion and was readily drawn into the political spheres of emergent Thonga and Zulu polities shifting between them as political pressures changed.' The roles played within the local society appear to be the norms and customs of the Thonga, whereas interaction with the world at large is conducted in the Zulu idiom[3]. As a people governed by the Zulu, it is expedient to pass as Zulu. It is interesting to note that while the men speak Zulu, it is extremely common to find the women conversing in Thonga (see also Webster, 1991).

Present day Maputaland now extends from Lake St Lucia in the south to the Mozambique border in the north. Its boundary is the Ubombo (Lebombo) Mountains on the west and the Indian Ocean on the east, and today comprises the Ubombo and Ingwavuma Magisterial districts. The area fell under the jurisdiction of the KwaZulu Government (the bantustan of the Zulu people). It now, subsequent to the first free and non-racial elections in South Africa held during April 1994, forms part of the KwaZulu-Natal provincial region.

The environment

Maputaland is composed of six ecological zones (Tinley and Van Riet, 1981): the mountains of the Lebombo (Ubombo) zone; the Pongola/Mkuze flood plain zone; the Sand forest zone; the Palm zone (Mosi flood plain); the Coastal lakes zone and the Coastal zone adjacent to the ocean.

All systems of agriculture described fall within the Coastal lakes zone. It is also this area that is targeted and/or has been enclosed to form nature preserves as a step towards the ultimate creation of a people-free national park.

The coastal zone is characterized by a chain of coastal lakes with forested sand dunes, undulating grasslands interspersed with savannah, thickets and forest. Swamp forests occur on many of the perennial bog drainage lines which enter the lakes. Six main systems of vegetation occur in the zone with a number of plant communities in each (Tinley and Van Riet, 1981; Van der Vliet and Begg, 1989):

- The estuarine habitat with its salt herb communities of rushes and succulents together with grass patches and mangrove trees.
- Fresh water aquatic and swamp communities with aquatic herbs, sedges, grass swamp and swamp forest in the swamp catchments, swamp drainage lines and along the boggy margins of lakes.

- o Open tree-less grasslands with dwarf shrub communities.
- o Savanna with tree and shrub species forming bush clumps and/or thickets in places.
- o Thickets comprising savanna; and forest species.
- o Mature evergreen forest with both tropical and temperate trees, that is sub-tropical.

Soils are mainly highly leached sands ranging from grey to white in colour, falling within the Fernwood form and exceeding 1.5 metres in depth. They may be wet or dry acid soils (MacVicar *et al.*, 1977). In the swamps there occurs a peaty black organic soil of the Champagne form. Both forms are utilized for agriculture A red Clovelly also occurs in patches but is not generally used for agriculture.

The area is frost free, with high humidity levels, and an average annual rainfall of approximately 1000 ml, occurring mainly during the summer months. Rainfall is often extremely heavy, resulting in a poor distribution. This distribution, when considered together with sandy soils of low water retention capacity, makes the area unsuited to agriculture.

Water utilized for domestic consumption is obtained from rivers and pans. Since the area is characterized by a high water table, shallow wells may be dug to obtain water for household consumption. However, water is often contaminated by human and animal waste. Infectious diseases account for a large proportion of morbidity in the area.

Infrastructure in the area is poor. As a result of its remoteness and Thonga rather than Zulu origin, the area has been neglected for political reasons. Matters have improved since the construction of a surfaced road in the 1980s as part of the South African government's defence strategy to counter the supposed threat of the Frelimo government of Mozambique (or according to a local headman (*induna*), the fruition of a recurring dream). Additionally, as a counter to attempts by the former South African government in 1982 to hand over the Ingwavuma district to Swaziland, the KwaZulu government reclaimed the 'Zuluness' of the area.

Sources of income

Whilst most of disposable income is earned through migrancy (Nattrass, 1977), other important sources are pensions, disability grants, salaries of KwaZulu civil servants mainly from the Manguzi Hospital, but with an ever increasing number in the employ of the KwaZulu Department (formerly Bureau) of Natural Resources. The only significant private sector employment is offered by the local supermarket. The informal sector is however growing rapidly. The local community and economy is dependent on cash derived elsewhere, rather than locally generated income.

Webster (1987) listed community perceptions of differences in status, conferred by alternate employment opportunities. Successful local subsistence agriculture was ranked after such categories as: self-employment (local shop-keeper); migrancy worker; locally employed (shop assistant) and pensioner. It is possible that agriculture would have rated higher before the intervention of nature conservation authorities with the concomitant threat of land alienation.

Since off-farm income-generating opportunities are lacking, the local population remains dependent on natural resource utilization. This includes fishing (an indigenous and ingenious system of fish traps realizes a sustainable harvest), gathering of wild fruits and vegetables, hunting (more so in the past since very little wildlife remains due to population pressure, poverty and the extinction of nearly all game animals in an attempt to eradicate the tsetse fly), pastoralism and agriculture. The use of wild medicinal plants plays an important role in the health of the communities. An urban market exists for medicinal plants harvested locally, as plants originating from Maputaland are attributed with exceptionally strong curative properties.

The harvest of the many wild fruits and vegetables is inextricably linked to agricultural systems; naturally occurring fruit trees left in fields (namely agro-forestry), colonizing species of annuals (weeds) harvested from fields as a relish, and fruit-bearing shrubs occurring in heavily utilized natural grassland. A hardy local breed of Nguni cattle (*Bos indicus*) is kept, but in comparison to Zulu society, milk products are of lesser importance.

Wild fruits and vegetables play an important part in the diets of the people. Traditional customs relating to the gathering of the first fruits are still retained. In the Chitamuzi area which lies adjacent to Lake Sibayi, residents still pay tribute to the local headman by contributing a portion of the first harvest of the fruits of the Forest milkberry (*Manilkara discolor*). The fruits of numerous other trees are relished, such as Wild custard-apple (*Annona senegalensis*), Wild medlar (*Vangueria infausta*), Marula (*Sclerocarya birrea*), Waterberry (*Syzygium cordatum*), Natal mahogany (*Trichilia emetica*), Spiny monkey orange (*Strychnos spinosa*), African mangosteen (*Garcinia livingstoneii*), Large num-num (*Carissa macrocarpa*) and Wild date palm (*Phoenix reclinata*) which is also tapped for palm wine (*isundu*).

Also harvested are wild spinach or the 'weeds' of gardens and fields. Numerous varieties such as the Pigweeds (*Amaranthus spp*) are utilized. Fruits of *Salacia krausii* are harvested from overutilized grassland.

Indigenous agricultural knowledge

The agriculture of the coastal zone includes forms of shifting cultivation uncommon to the rest of KwaZulu. Whilst not all the agriculture incorporates 'field rotation', generally all farmers employ a form of shifting cultivation. Much of the agriculture is opportunistic in the sense that the utilization of different habitats (from dryland to wetland) will be a function of the rainfall pattern occurring during the planting season. This view concurs with Richards (1985), who refers to shifting cultivation as neither a system nor a stage in agricultural development, but as part of a 'tool kit' for land management.

Taylor (1988) described three main systems of agriculture, namely gardening, dryland cropping and swamp farming. All households are likely to have a field and some form of homestead garden; only those with access to a wetland will have a swamp farm or a wetland garden. Gardens may take the form of a homestead garden which will include both annuals and perennials, or alternatively an individual or communal vegetable garden near a water source.

Gardens

Gardens can make an important contribution to the household diet (Cleveland and Soleri, 1991). Whilst the average individual and communal vegetable gardens supplement and add variety to the diet, their contribution to household nutrition is not as important as a swamp farm or dryland cropping. However, depending on the scale of operation and the plants cultivated, the homestead garden can be a critical source of food. This is especially applicable in the years of drought, when dryland cropping fails. The homestead garden can include citrus (*Citrus spp*), mango (*Mangifera indica*), pawpaw (*Carica papaya*), guava (*Psidium guajava*), cashew (*Anacardium occidentale*), granadilla and guavadilla (*Passiflora spp*), pineapple (*Ananas comosus*) as well as maize (*Zea mays*), rice (both *Oryza sativa* and *Oryza glabberima*), pumpkin (*Cucurbita spp*), sweet potato (*Ipomoea batatas*), taro (*Colocasia esculenta*), numerous vegetables such as cabbage, tomato, spinach, onion and brinjal interspersed with some of the indigenous fruit trees mentioned earlier. The homestead gardens can be located adjacent to wetlands, thus moving from dryland conditions around the houses to moist soil regimes as one moves further away. In drier locations, apart from a few fruit trees, some vegetables and sweet potatoes may be grown. Since homesteads are scattered throughout the countryside, the most productive gardens are those of the former.

Some innovative farmers rotate the cattle kraal (enclosure) and plant vegetables in this highly fertile environment. Alternatively, the kraal may be situated above the garden (or field) so that nutrients are washed into the garden below. In these cases water is often carried to the garden to supplement rainfall.

A communal vegetable garden is farmed jointly by a group of women, each of whom has a designated area for her own use. This form of agriculture has been encouraged by the state agricultural departments and non-governmental organizations (NGOs), to facilitate agricultural extension efforts.

Dryland cropping

Dryland cropping takes place in the highly leached sandy soils which are acid and low in fertility. It is a semi-intensive to intensive rain-fed system of agriculture, dependent on an erratic rainfall. The effective rainfall in these soils of exceedingly low moisture-retention capacities is low, despite the relatively high average annual rainfall. It is the rainfall distribution and poor soils that make the area marginal for crop production.

Land-races are the predominant varieties used in the area. Modern varieties are used when local varieties are in short supply following successive years of drought, or alternatively, when they occupy a specific niche in a cropping system. For example hybrid maize is planted because of its early maturity.

Cultivation is normally done with a hand-hoe. Where possible the afforested areas are chosen. Tree stumps make it impossible to use draught tillage. A few farmers use oxen in the more open areas. Since population pressure is relatively low, land is available for shifting cultivation. Forests are now protected and so it is generally rested areas that are re-cleared. In forests and/or thickets, trees are chopped down, vegetation slashed, allowed to dry and then burnt. Shifting cultivation has led to a significant reduction in afforested areas. In virgin grassland, or where grass is long, the area is burnt and then dug by hoe. The burnt tufts are then placed in heaps in or adjacent to the field, and are either left to rot or are burnt. Many farmers have realized the need for soil organic matter and so prefer not to burn. Hoed grass tufts may also be left to rot where they are removed, with planting taking place around them.

After sufficient rainfall has fallen, planting takes place. The area planted will be a function of the rainfall received and of labour availability. Farmers will continue preparing land and planting provided soil moisture is adequate, constituting a form of staggered cultivation. Variations in practices occur. Thus for example, a common and interesting procedure is to broadcast sorghum (*Sorghum bicolor*) and millet (*Pennisetum americanum*)[4] seeds in the field prior to cultivation. Linked to this is the practice of planting the seed of a small indeterminate cowpea (*Vigna unguiculata*) variety between growing grass tufts. After germination of the cowpeas, the grass is hoed out and maize planted. A field could then have millet, sorghum, maize and cowpeas growing simultaneously. Cassava (*Manihot esculenta)* and pumpkins (*Curcubita spp)* may also be planted.

Every household plants a crop of maize. Maize is generally planted in holes (3 to 4 kernels in each hole) separated by one metre and scattered throughout the field. The combination of maize, sorghum, millet and cowpeas in a single field is a form of multiple cropping which allows a harvest over an extended period of time. Since wild fruit trees such as marula, Natal mahogany and monkey orange are often left in fields, the combination of annual crops and perennial trees constitutes a form of agroforestry (Taylor, 1991). Pumpkin is often planted with maize. In addition to the pumpkin itself, the young shoots are picked and eaten as a spinach. Groundnut (*Arachis hypogaea*), the second most commonly grown field crop, is planted as a monoculture, as is the Bambara groundnut or Jugo bean (*Voandzeia subterranea*). Whereas groundnut land-races are preferred, a number of farmers have depleted their seed reserves to such an extent that less suitable modern varieties are purchased. Jugo beans are popular but there is no external source of seeds, and farmers, once they have consumed their own supplies or experienced successive crop failures, have nothing to replant. Neither groundnuts nor jugo beans are planted in the same field in consecutive years.

A larger variety of cowpea is planted where seed is available. This indeterminate variety planted in holes one metre apart, soon covers the whole field and serves as a mulch for the maize with which it is usually planted. This variety gives an extended harvest over a number of months. The young shoots are eaten as a relish. Where sweet potatoes are planted in fields, a ridge and furrow system is used with cuttings planted on the ridges. A number of local varieties are planted ranging from early yielding to late yielding varieties. Cassava is planted by many households, but is often only harvested in times of drought as it is regarded as a 'poor man's crop'. As the major crop of Mozambique, it would have been more widely grown in Maputaland in the past.

Cleared areas previously afforested are utilized for approximately 10 years before being rested for 3 to 4 years. Cleared grasslands are cropped for 3 to 5 years before being rested for 2 to 3 years. This period of fallow replenishes soil nutrients but impacts on the natural vegetation as new areas are cleared.

Since the majority of able-bodied men are migrants, labour is a constraint. Work parties known as *illima* are organized where food and drink is exchanged for the labour required to complete a portion of a field. This system of reciprocity allows larger areas to be cultivated and in fact the very poor may well be dependent on the system (Webster, 1991). Taylor (1988) mentions another system known as *isinenene*, which is no longer practised. Labour was offered to complete an entire field on the understanding that a post-harvest party would be held.

Swamp farming

The importance of swamp habitats to the food security of the Maputaland household must be understood in the context of an area marginal for agricultural production. In years of drought the swamp farm never fails to realize a yield or in the words of an old man, '*umnotho wethu amadobo*' (the swamp forests are our fertility). But it is these areas that are ecologically critical for the maintenance of the Kosi Lake system, and access is contested by both farmer and conservationist.

In some of the swampy areas, a system of mound farming is practised. A rectangular drainage ditch is dug forming raised rectangular beds (mounds) which incorporate large amounts of organic matter and peaty soil. A single bed or series of beds may be constructed. Sweet potatoes, taro and sugar cane (*Saccharum spp)* are the main crops planted in these beds but vegetables may also be cultivated. In times of drought these beds may be used for planting maize. The beds are constructed in wetlands inhabited by a herb/sedge/grass sedge community. Some of these areas could have been swamp forests previously. Mounds are often also constructed in the wet sandy areas where farmers lack access to the swamps. The same principles as above are adopted.

Swamp farming itself refers to the utilization of wetlands involving deforestation of the tree vegetation (Taylor, 1988). Swamp forests are of the equatorial-rainforest type but include some of more temperate forest species (Tinley and Van Riet, 1981). Species present could include the Powder-puff tree *(Barringtonia racemosa)*, Wild frangipani *(Voacanga thoursii)*, Wild Poplar *(Macaranga capensis)*, Swamp fig *(Ficus trichopoda)*, Waterberry *(Syzygium cordatum)* and others. It is a form of shifting cultivation with areas utilized from 15 to 40 years before allowing a fallow period. The upper limit figure of 40 years is an estimate as the area described still continues under banana production (see Taylor, 1988). However some farmers claim that swamp soils may be utilized permanently provided they are correctly managed.

In preparing a swamp farm, deforestation of an area takes place. Trees are felled and undergrowth is cleared. The area is left to dry out and is then burnt. Some farmers leave the larger trees intact, whilst others chop trees at chest height which will enable rapid regrowth for the fallow period. Sometimes felling and slashing of undergrowth is done some months before cultivation and left to rot for 3 to 4 months before planting. Planting into this mulch then takes place. According to farmers, a swamp farm which is rested or remains unweeded will regenerate after a fallow period of 3 to 4 years. It appears that most of the existing swamp forest has been utilized previously. For example, an area said to have been last utilized in the 1940s, appears to be 'climax' forest in the present time. It can therefore be inferred that swamp farming is an age-old technique, and the Kosi Lake system as it exists today is at least partially a product of the activities of farmers. (The effects of deforestation on species diversity remains unknown at this point in time as no studies or monitoring have been done, and thus systems degradation, if any, remains unmeasured.)

A newly planted swamp farm could look as follows: bananas and to a lesser extent plantains (*Musa spp)* will be planted 3 metres apart, interspersed with taro and sweet potatoes. Rice could be planted in patches, as could cabbages and other vegetables. Maize and pumpkins may be planted on the periphery, moving up the adjacent bank into sandy soils. Sugar cane will also be planted. Indigenous moisture-tolerant trees such as the Waterberry will remain as part of the system. Whilst at first sight the planting appears to be completely random, areas will be matched to plant requirements. Once again planting is opportunistic, in the sense that depending on the rainfall, planting might take place in fields or swamp farms. Thus in the dry years maize would be planted in a swamp, whereas in the wet years conditions would be unfavourable for staple production, but suitable for taro and sweet potato. Since the sandy soils are the preferred areas for maize cultivation, the use of the swamps is a hedge against unfavourable climatic conditions, as they are both nutrient and water 'oases' during drought periods.

Soil moisture is controlled by drains and ditches. In wet seasons the swamps become waterlogged and it is necessary to drain excess water. Deep drains following the main water course are dug and will eventually enter a river/stream or lake. Since these drains are fairly far apart the soil does not dry out. Alternative areas may be irrigated if necessary, by blocking one or more of the drains thereby artificially replicating the natural flooding regimes. Since the swamp farms are periodically flooded, soil nutrients are replenished by nutrient-laden water flowing through the system. This could substantiate the claims of many farmers that the soil never tires.

Although a mosaic of crops is planted initially, bananas eventually colonize the whole area if not prevented from doing so, but if left the whole area will in time revert back to forest. This is the main reason for opening up new areas. The importance of banana production is that, without exception, it is the one crop which always generates a surplus.

The clash of knowledge systems

Whilst swamp farming does occur outside the protected areas, many of the farming systems described fall within the now declared Kosi Bay Nature Reserve which is under the KwaZulu Department of Natural Resources. This conservation body has planned a 'U-shaped' National Park which will follow the Swaziland border in the west, run along the Mozambique border in the north and down to St Lucia in the east.

The Kosi Bay Nature Reserve proclaimed in 1988 is an area of approximately 11 000 hectares. All cultivation inside the reserve has been banned and people have been told that they will have to move (AFRA, 1990; CORD, 1991a). Although some households have refused to move, many families have had no alternative but to relocate themselves outside the Kosi Bay reserve boundaries. These removals have not been forced in the classical sense, that is the demolition of people's homes; but fencing has reduced mobility, agriculture is prevented, hippopotamuses destroy crops and people no longer feel free in their own homes.

Swamp forest over-exploitation is cited by conservationists as one of main reasons for preventing agriculture; the effect of cultivation on water-flow into the lakes is said to be detrimental to the system as swamp forests are obviously nutrient and detritus filters. There is no empirical evidence to suggest that this has caused degradation. The late Chief Mzimba Tembe suggested that the increased flow of water kept Kosi Mouth open and therefore salinization levels down (M. Tembe: personal communication 1985). Farmers on the other hand claim that since swamp forest cultivation has occurred for decades without destroying the ecosystem of the lake, it cannot be detrimental. This is confirmed by Begg (1980: 372) who claims, 'the present day environmental condition of the Kosi System can be regarded as highly satisfactory'. He does however see 'slash and burn agriculture' as a threat to the system.

This clash between indigenous agriculture and a Western conservationist ethic is more than competition for conflicting land use. It is the clash of knowledge systems – between the universality of a Western conservation ideology with its concept of sustainable resource utilization, and indigenous knowledge with local needs and aspirations which necessitate land usage. But the Western conservationist ethic within the context of the African continent, is underpinned by a preservationist legacy, a relic of the colonial era. It is sustainable land use as rhetoric rather than action, with Nature Conservation personnel adopting paramilitary attire, marching, saluting, carrying arms, culling (often necessary but still an opportunity to hunt as in the Africa of old), and the habit of being white veterans of wars against indigenous peoples.

In attempting to conserve the area for posterity, the Department adopts an extremely paternalistic attitude to local people. The implication that local farmers cannot manage their environment but that outsiders can, is a serious indictment of the Department's (formerly Bureau) claim to work with people. According to CORD(1991b), 'Consultation by the Bureau's conservation programmes is with local tribal authorities and pays scant attention to indigenous knowledge in the maintenance of a unique ecosystem. The extent of current and ongoing dispossession in this region can be gauged by the fact that almost 80 per cent of the Bureau of Natural Resources annual budget is spent on conserving and controlling an environment which is only unique because it has not been mismanaged and abused by the indigenous population'.

In the eyes of the local population the Department is more concerned with animals than it is with people. Nature reserves are seen to be the preserves of the rich and – in South African society – white. The rural poor are once more being discriminated against but this time in their own back-yards. There is little wonder then that the Thonga Independence Party which surfaced in 1985 in unsuccessful opposition to the KwaZulu Government, should re-emerge in the 1990s in sympathy with the African National Congress. It is the conservation issue more than any other that has united people against an administration which has offered them very little. A local resident went as far as to say that millet, the traditional cereal crop of the Thonga is becoming more popular relative to maize but this is not substantiated.

Recent studies refer to evidence which indicates that farmers have lived in parts of the Southern African sub-continent for two thousand years (Granger *et al.*, 1985) and 'that it is more appropriate to assume that the present flora of Southern Africa has been formed at the hand of man over many centuries than to trust in the pre-colonial wilderness model.' Therefore, if most of the swamp forests have been farmed previously, then people's intervention is part of the natural cycle. The question that arises is how to incorporate conservation in an appropriate and meaningful development framework which reduces the incidence of harmful practice or minimizes the need to utilize environmentally sensitive areas. Limiting rather than preventing access is required. Even if pressure on the swamp forests has increased over the previous decade because of population increases, there is, with our present level of knowledge of the dynamics of the system, no justification for making the swamp forests a no-access area for agriculture. In fact, steps have been proposed by the Department to limit swamp forest cultivation outside the reserves. It is somewhat paradoxical that a proven and successful system of no external input agriculture namely swamp farming, is perceived as unsustainable by a conservation body that claims to work with local people.

This all has serious repercussions on farming. Since most farmers are sub-subsistence farmers – it has been estimated that average yields for maize producing households in Maputaland is approximately 500 kilograms which is 25 per cent less than household subsistence requirements (VARA, 1989) – wetland utilization is often the only bulwark against destitution. Maize yields measured for the 1993–4 planting season, a season regarded by farmers as less than satisfactory, varied from 0.5 to 1.5 tons per hectare utilizing no inputs. It must be noted that often less than a hectare of land is cultivated.

A strategy of confusion reigns

Maputaland has been subjected to numerous studies, all of which point to the need to develop small scale agriculture with approaches that vary from 'trench farming' (Tinley and Van Riet, 1981) to the provision of inputs and farmer support services (VARA, 1989). There has been no recognition of the ability of the low resource (resource poor) farmers to survive successfully in such a poor environment, no acknowledgment of farming innovations, nor cognisance taken of their ability to farm a diverse range of habitats. On the contrary local knowledge has been deliberately undermined in the South African context (Taylor, 1993).

Numerous costly and failed Maputaland interventions can be cited, including the coconut plantation near Kosi Bay, the Mjindi Project on the Pongola River (a large irrigation scheme which has meant settling and resettling farmers) and the KwaZulu forestry plantations; not to mention the non-governmental organizations (NGOs), who battle with one another over questions of representivity, ideology and delivery. Therefore a new approach is essential. First and foremost, people's inter-relationship with the environment needs to be better understood and accepted. The concept of pristine wilderness must be replaced by the realization that the environment has been, and will continue to be shaped by people. In areas such as Maputaland where people have lived for centuries, it must be understood that agriculture will remain part and parcel of areas with conservation potential.

It is therefore necessary to identify the research priorities, undertake the appropriate research, and formulate strategies for the appropriate management of each designated area. Saterson (1990) lists a number of criteria – falling within the ecological, administrative and economic spheres – that should be considered when evaluating the need to protect biological diversity.

- Level of species endemism and richness of habitat.
- Degree of human threat and vulnerability of species or ecosystem.
- Economic and ecological importance to local human needs.
- Importance of habitat to maintaining diversity elsewhere.
- Similar areas protected elsewhere.
- Political and economic factors that could help or hinder successful protection.

This framework attempts to identify the minimum critical area needed to sustain plant and animal diversity in the face of population pressures. It is a management plan which recognizes the need for the integration of conservation with alternative land uses. However, the process places inadequate emphasis on local fears and aspirations, failing to analyse the decision-making process itself, differences in stakeholder priorities and how implementation can take place in a contested landscape. There is an implicit faith in outside experts and expertise, whose assumptions have so frequently been wrong. The multitude of development failures in Maputaland give credence to this statement.

There are considerable limits to even the best of planning, and whilst environmental impact and changes in species composition should be monitored, relatively little is known of the effects on long-term vegetation changes and systems resilience in the African environment. In the context of South Africa's transition and social transformation, it is critical that local people perceive conservation at best, as beneficial, or, at worst as relatively benign. Enclosing the commons for tourism by the rich, is nothing short of a recipe for disaster, particularly since the material benefits of imposed conservation schemes seldom accrue to those most directly affected by them.

Maputaland, considered regionally, is part of the coastal strip running into Mozambique with the same or similar flora and fauna but becoming more tropical as one moves northwards. The need for the conservation of biological diversity requires a regional perspective. The special nature of the 'pristine wilderness of Maputaland' has perhaps been overemphasized. However, there is a way out with regard to swamp forest utilization.

- Define together with local representatives priority areas for protection.
- Distinguish between preservation (perhaps no other land use other than tourism) and conservation areas, and define acceptable and sustainable forms of land utilization.
- Examine existing land use patterns, and allow present use to continue where possible with the agreement that no further major habitat change takes place without further negotiations (including tourism infrastructure). Looking towards belts of agriculture interspersed with forest appears to be a meaningful compromise.
- Offer compensation and/or off-farm income-generating activities.
- Allow affected homesteads to remain and make them part of a tourism plan.
- Facilitate the emergence of representative farmer/rural organizations which can ensure that the necessary controls are acceptable and will be maintained.

The history of conservation in South Africa has been one of dispossession. Whilst the KwaZulu Government's White Paper (1992) on a development policy for Maputaland sets out clearly the need for 'balanced development', the implementation leaves much to be desired. However, in the current political climate, external pressures are forcing the nature conservation authorities to negotiate conservation priorities with community-based organizations, previously regarded as the 'enemy'.

Discussion

This chapter has described systems of agriculture under threat. Although these have been described under the categories of gardening, dryland cropping and swamp farming it is apparent that the three are not discrete systems but merge and separate depending on the environmental conditions at any point in time. The three systems are, then, the farmer's 'tool kit' through which household food security is ensured in the face of climatic uncertainties. Agriculture thus refers to dynamic systems which respond both to weather patterns and the political and economic macroenvironment.

Farmers are still largely dependent on land-races, but have eclectically incorporated modern cultivars, yet have received little recognition for their abilities. There has clearly been a failure to recognize that farmers have survived for centuries in Maputaland, and deserve (apart from their basic rights to be part of any development process) to be part of a solution which will necessarily include both agriculture and conservation. Social systems have changed and with them the social institutions which have governed access to local resources. Yet it is a fallacy to believe that Maputaland is a free-for-all.

Rhoades (1990) identified four major stages through which agricultural research has passed or will pass; namely production, economics, ecology and social institutions. It is the latter that is critical in the Maputaland context. For example, in 1986 the Thuthukani Farmers Co-operative was established as the umbrella body for some 34 local organizations. It was established with the following objectives in mind, as:

- a source of information and inputs
- a marketing channel for farm products
- a co-ordinating body for farming activities, such as production and training
- a means to mobilizing/organizing farmers
- a representative body of farmers with a mandate to lobby and advocate on their behalf, and to enter into negotiations with the public and private sectors.

Following the eras of colonization and apartheid, South African society requires a form of social reconstruction which transforms the institutions of state and civil society. This will include both agriculture and conservation in new forms of management and control, incorporating government, the private sector and local institutions. In the Maputaland context, a failure to achieve a degree of agricultural sustainability together with a realistic conservation policy, is a scenario for the failure of both.

Conclusion

Agriculture in Maputaland has allowed the survival of people under difficult circumstances. These agricultural practices have been dynamic, innovative and opportunistic and have been built on local environmental knowledge over centuries of land use. This has allowed farmers to take advantage of a variety of habitats, at different times, depending on their social circumstances.

Swamp forest utilization has replaced a tree community, with primarily a banana crop which constitutes a successful form of no external input agriculture. There is the danger that unlimited use, resulting from population increases could jeopardize the Kosi Lake system. Yet at the same, time we are concerned here with an area that has never been a pristine wilderness, but has been shaped through centuries of use.

Sustainable development in Maputaland must include both successful agriculture and conservation. However, it is only through strong and appropriate local institutions, that can support farmer aspirations, meet rural needs and revalorize the forms of indigenous knowledge necessary for sustainable agriculture, that this is likely to achieved. The best guarantor of a workable environmental conservation policy is one of co-existence with local agricultural efforts.

The word '*umnotho*' has a double meaning; in a narrow and particularistic sense it means fertility, whereas when applied more broadly, it refers to wealth and prosperity. The conservation of the environment is intertwined with the conservation of the culture, identity and prosperity of the human agents that have shaped it. It is in this sense that the swamp forests are 'wealth' for both agriculture and conservation.

14. Local-level experimentation with social organization and management of self-reliant agricultural development: the case of gender in Ara, Nigeria

D. MICHAEL WARREN and MARY S. WARREN

Abstract

THE CASE-STUDY of Ara, Nigeria explores the gender-based issues that emerged in a one-week development planning workshop for leaders of the numerous indigenous associations of Ara. The issues were addressed by the 110 participants in the workshop as objectives in an 18-month development plan. The objectives ranged from structural changes in the organization of community development associations to the design of new institutions (such as a day care and community bank intended to eliminate key constraints faced by citizens of Ara, particularly the working women), as well as the search for improved appropriate technologies for the processing of important crops such as oil palm and cassava. The case-study indicates the willingness of citizens of a community to explore and cope with the problems and opportunities that it faces, in innovative ways.

Introduction

Every community faces a universal constant, the dynamics of changing sets of circumstances that represent both challenges to, and opportunities for, that community. The initial step for a community in addressing situations of change is the identification and prioritization of the range of locally perceived problems and opportunities that change represents. Frequently the discussion takes place in groups, where the differential perception of problems and opportunities varies according to the composition of the groups and the special interests represented. Community-based groups range from the informal – such as women who happen to be processing palm oil in the same locality at the same time – to the formal, such as a community development council. To understand the heterogeneous nature of the community problem-identification process, one must understand the variety of groups within a community as well as their structure, composition, and functions. This case-study deals with the changing calculus of community decision-making when the composition of the primary associations dealing with community development was changed to include a higher representation of women.

Ara is a Yoruba community of 1000 households, with a population of about 10 000, located in the tropical rainforest about 15 miles from Oshogbo, the capital of Osun State in Nigeria. The community is an ancient town that traces its origin to the founder, Orira, who migrated from Ile-Ife, possibly as early as the tenth century. Orira was a son of Aranfe, one of the sons of Olofin Oduduwa, the forefather of the Yoruba people. About 90 per cent of the adult inhabitants of Ara are engaged in full-time and part-time farming. The predominant farming system is a perennial mixed-plantation system of trees, bearing cash crops such as kola, cocoa, oil palm, coffee, coconut, cashew, citrus, guava, rubber, and mango. Of growing importance is the farming system based on the biennial and annual mixed cropping of arable crops such as yam, plantains/banana, cassava, cocoyam, maize, papaya, sweet potato, beans, groundnuts, and a wide variety of vegetables such as okro, tomatoes, egusi melon, pepper, and numerous types of leafy greens. Although these food crops were grown on a subsistence basis in the past, many of them have now assumed the status of cash crops.

Ara is particularly well known for its production of high quality palm oil, which has been produced for many generations by women who use a very innovative traditional process that is also very time-consuming and laborious. Although the women control the palm-oil production process, they must obtain the palm nuts from their husbands and other male farmers, since traditionally trees are owned and controlled by men. The women palm-oil producers must provide a share of the palm oil to the owner of the palm tree from which the nuts came, but they have rights to all profits made from the sale of their share of the palm oil as well as to the sale of any byproducts such as the palm fibres, shells, and palm kernels.

The actual sales of the palm products, retained by the women producers, represent significant income for the community as several of the products are of high value. Although the palm oil is the highest value product, the palm fibres, the shells from the inner nut, and the palm nut kernels are also in demand. The palm fibres and shells represent high-quality smokeless fuel used for cooking, whereas the palm nut kernels can be processed into palm kernel oil – the basic ingredient for the production of traditional soap – and palm kernel cake, an important ingredient for the production of animal feeds. The palm fibres and inner palm nuts are separated from the palm oil in the initial traditional production step. After the inner palm nuts are dried, women and children from the household crack them individually by hand, the broken shells being separated from the kernels. Since the community has no experience with any traditional process for the production of palm kernel oil and cake, the women have been forced to sell the palm kernels to outsiders at very low prices.

Although Ara is a relatively small town, like most communities it has a wide range of formally structured organizations, most of which have among their functions that of community development. The range of associations includes social clubs, religious groups, occupational associations, and community-wide associations such as the Ara Traditional Council, the Ara Descendants Union and the Ara Development Council. The Ara Traditional Council comprises the Alara, the *Oba* or king of Ara and its 36 surrounding towns and villages, and his 23 chiefs, of whom several are women. Only one of these towns has a leader with the rank of *oba*, while seven of the towns have leaders with the lower rank of *baale*. The numerous small farming villages have headmen as their leaders.

The Ara Descendants Union (ADU) was established in 1947 as a hometown association intended to unite the growing number of Ara citizens who migrated to numerous other communities both within Nigeria and other countries, particularly the Ivory Coast where about 500 citizens currently live and work. There are 19 active branches of the ADU, several of which have been important forces in shaping and supporting Ara's development.

The Ara Development Council (ADC), established in 1949, is the most recent of a series of community-wide development associations that can be traced back through oral history to more than 100 years ago. The ADC and its precursors have a proud history of self-reliant development projects that include the construction of the first roads and bridges linking Ara to surrounding towns and other Yoruba traditional states early in the twentieth century, the Baptist Day (primary) School (1933), the Customary Court (1950), the Town Hall (1951), the Ara Postal Agency (1951), the Farmers' Co-operative Society (1950), a maternity clinic (1954) and dispensary (1956), and more recently the extension of both pipeborne water (1988) and electricity (1990) to Ara. The composition of the council has been based on representatives of the most prominent households in the community. As a patrilineal and patriarchal Yoruba community, these representatives were usually males.

Formal leadership roles have typically been filled by males in Yoruba society. This does not mean, however, that females have not played significant roles in shaping community opinion. Several Yoruba communities have had female *obas*, and all Yoruba communities have a few female chiefs, such as the *iyaloja*, the chief in charge of the marketplace. Most of the formal associations in Ara include females and some associations are predominately or exclusively female. These include social clubs, religious groups, and the Ara Branch of the Better Life for Rural Women Programme that was initiated as a national programme by Miriam Babangida, wife of the former head of state, with the Ara branch being established in 1989. Although men have traditionally played the primary role in production agriculture, women have been assuming a growing role in this area during the past few decades. Women play a predominant role in processing many agricultural products such as palm oil, *gari* flour from cassava, and *elubo* flour from yams.

In August 1991, a one-week Ara community leaders' management and development planning workshop was conducted by Dr. D. M. Warren, an American married since 1967 to Mary Warren, a citizen of Ara. D. M. Warren, installed as the *Atunluse* of Ara in 1990 – the chief in charge of community development, had spent many years involved in development projects in Africa designed to improve development planning for local administrative units such as District Councils. This workshop was different as it was conducted in the Yoruba language. Although the workshop was supported by the officials of the Egbedore local government, the actual workshop was sponsored by community-wide development associations such as the Ara Traditional Council, the ADU and the ADC. There was strong interest among community leaders in being exposed to a variety of management and development-planning skills that were typically covered in local government workshops that Warren had conducted in places like Ghana and Zambia.

It was decided within the community to include in the workshop as many leaders of Ara associations as might want to devote a week's time. The participants included the *Alara* and his chiefs, as well as officers from the Ara Development Council, the Ara Descendants Union, women's groups such as the Ara Branch of the Better Life for Rural Women Programme, youth groups, occupational associations, and religious groups. The final composition of the participants was very heterogeneous, including numerous women and young people.

One of the functions of the workshop was to list and prioritize as many problems and constraints to development in Ara that participants could identify. Given the unusually broad representation of interests in the participants, the workshop elicited a wide array of viewpoints. This chapter will focus on the development constraints that represented gender concerns and how these concerns have been turned into innovative approaches to social organization and management of issues that relate directly to self-reliant approaches to agricultural development in Ara.

Gender issues that emerged in the workshop included the under-representation of women on community-level decision-making associations such as the Ara Development Council, inadequate long-term supply of palm nuts due to ageing trees and lack of access to hybrid oil palm seedlings, credit limitations faced frequently by women, child care constraints for women involved in agricultural activities, technology limitations in the processing of palm oil and *gari*, and marketing constraints for agricultural products such as palm oil and palm kernels.

Workshop participants developed an 18-month development plan that addressed the gender-based issues identified as well as numerous other issues faced by the community of Ara. A decision was made to conduct an annual two-day evaluative workshop every December to discuss the achievements made on the current development plan and to design the new annual plan.

Participants recommended significant structural changes in the Ara Development Council. For the first time, two-year terms were set for officers who would be elected in a general community gathering. It was decided to alter the structure of representation away from extended families to one that was associational in nature. Representatives were to be chosen by their respective constituencies from women's associations, occupational associations, co-operative societies, the Christian community, the Muslim community, commercial associations, the Ara Descendants Union, the PTAs, the Traditional Council, the social clubs, the *obas* and *baales* representing towns under the *Alara* of Ara, and the smaller towns and villages led by headmen. The ADC was to meet monthly at times announced well in advance. Any Ara citizen was to have the right to attend any of the meetings as a non-voting participant.

The 1993 evaluative workshop in Ara discussed the effectiveness of the structure of the ADC as implemented in 1991. It was decided that a third model should be tried during the 1994–5 period, single representatives being chosen by the major types of associations along with single representatives of the major extended family households in Ara. It was felt that this third approach, combining elements of the original structure and the 1991–3 approach, would help to enhance communications and information flow between a larger proportion of the community and the Ara Development Council. In 1991 and 1992 committees were set up to oversee the planning and implementation of numerous projects laid out on the 1991–2 development plant. A concerted effort was made to assure that women were included on each committee. The committees established to design an improved market and to initiate improved approaches to the production of palm oil were chaired by women. In terms of individual and institutional capacity-building for leadership, management, and development skills within the community, it was decided in 1993 to have annual terms for officers of committees so more citizens could gain these important skills.

One of the key constraints discussed in the 1991 workshop was the lack of financial fluidity at times when agricultural products were not available for sale. This seemed to affect women particularly. Now that universal primary education is in place in Nigeria, all Ara families are faced with new financial outlays for school uniforms and fees, books, and other related costs. For large families the schooling of children can represent a major burden. With pipeborne water and electricity becoming available in Ara during the past several years, it is now realized that these new luxuries also come with utility bills that represent another financial burden. Both women and men explained that they can be caught in a financial situation aggravated by unexpected expenses due to such things as illness in the family. When short-term loans are not available from other family members and friends, the final option available is to take a loan from a money lender from one of the nearby larger towns. The terms of these loans are extremely unfavourable. The money lenders prefer repayment in kind, such as a five-gallon container of palm oil, access to all of the fruit on a given number of fruit trees (e.g. kola, orange, oil palm) for one or more seasons, or pawning one's farm for a given number of cropping seasons. It was discovered in the workshop that some villagers, desperate for immediate cash, have even sold high-value tropical hardwood trees for as little as Naira 20–100, a practice now halted due to action taken at the workshop (the 1991 exchange rate was about N20 = US$1). The return on the loan to the money lender ranges upward from about 500 per cent.

Although Ara has had a branch of one of the major Nigerian commercial banks since 1988, the operating guidelines governing the issuance of credit follows standard commercial banking practices. There are minimum limits for loans that exceed the amounts that many women need to take them through a financial crisis until the next palm nut harvest when new palm oil is available for sale. Collateral is also required in forms usually not easily available to many women. It was discovered that a very innovative programme for addressing these types of problems (that are common in many Nigerian communities) had been established by the government of Nigeria. The Community Bank programme provided mechanisms allowing communities to raise their own share capital (with a minimum of N250,000) from their own citizens (with no single citizen acquiring more than 5 per cent of the total share capital), to build and operate their own bank, and to be duly registered and monitored by government regulatory agencies. The operating principles of the community banks include the provision of very small loans at short notice without the standard requirement of collateral. It was decided that if Ara could achieve the major objective of establishing the Ara Community Bank, the small-scale credit needs of its citizens would be well met. By the end of 1993 the community had raised the minimum share capital, obtained the initial licence from the government, and had secured sufficient donations from within the community to construct the new bank building next to the *Alara*'s palace up to the roofing stage.

Child care has traditionally been assumed by women and the siblings of infants and young children. With universal primary-school education, virtually all of the school-age children within a household are at school for extended periods of time during the school week. For households without capable elderly women present to care for infants and pre-school-aged children, women working in production, agriculture and in the processing of agricultural products such as palm oil and *gari*, must also assume the additional burden of infant care. Participants in the 1991 workshop noted that many of the urban centres had established day care centres. Discussion of these issues led to the objective of establishing the Ara Community Day Care facility. It was decided to rehabilitate the former maternity building that had been abandoned and overgrown by bush after the new maternity facility had been opened in 1987. Numerous community fund-raising activities took place. In the summer of 1993 the Ara Community Day Care was formally opened with 42 three- and four-year old children being cared for by two day-care facilitators. The monthly charges for a child are minimal (N30 per month), but still sufficient to cover the salaries of the facilitators with a small additional amount accruing in a bank account to cover maintenance costs and future capital development costs. Donations have covered the complete rehabilitation of the abandoned building including masonry, carpentry, and painting, extension of water lines for potable water for a shower and kitchen sinks, a kerosene stove, small tables and benches, beds for nap times, painting of pictures with their English names on the outside and inside walls, chalk boards, and hand-constructed playground equipment purchased in Ibadan and transported to Ara. The facilitators provide lunches for the children. Daily instruction in basic English and Yoruba language and in arithmetic is expected to provide a strong foundation for the children once they enter primary school at 5 years of age. The community has now dissolved the Day Care Planning Committee and replaced it with an Ara Day Care Advisory Board and a PTA. The working mothers are confident that their small children are in excellent care while they are on the farm or busy processing agricultural products.

The production of both palm oil and *gari* has been carried out in Ara for generations based on highly effective traditional technologies that are, however, both time-consuming and very laborious. Workshop participants were aware that new companies in major urban centres such as Ibadan were producing appropriate technology machines – some already in place in nearby communities – that greatly reduced the time and labour involved in production of palm oil and *gari*. It was also known that the national and state offices of the Better Life for Rural Women programme had supported some women's associations in acquiring this new technology. It was decided in the workshop to establish a committee to explore the various possibilities for making palm oil and *gari* production more efficient. The Ara Development Council and the Ara Branch of the Better Life for Rural Women programme submitted joint mini-proposals for funding for these machines to various small-scale project funds available through the ambassadors of major embassies in Lagos. Although the initial response from some embassies was disappointing, the US Embassy expressed preliminary interest and sent a delegation to Ara to explore the situation. It was agreed that the US Ambassador, through the ambassador's small-scale project fund, would cover the costs of acquiring the appropriate technology machines and engines needed for the women's project. By the end of 1993, the community had received from the manufacturer in Ibadan four machines for *gari* production, five machines for production of palm oil and palm kernel oil and cake, a three-phase electric motor, two diesel engines, and a petroleum engine. This was most exciting as it provided the possibilities of larger-scale production of *gari* and palm oil, as well as – for the first time in Ara's history – the actual production of palm kernel oil and cake, opening the new opportunity for future expansion into production of traditional soap and animal feed.

Although the major costs of procurement of the machines and engines had been covered by the US Ambassador's small-scale project fund, the community faced the challenge of construction of a facility to house them. Again, numerous pathways were pursued to raise the necessary funds. The Ara Descendants Union in Abidjan, Côte d'Ivoire, was most responsive to the special appeal by the women for financial assistance. By the end of 1993 a large building had been constructed with cement floors thick enough to withstand the vibrations of the machines, pipe-borne water, and the partial extension of electricity from the main road to the food processing facility.

The facility included an office and a storage room so products could be stored when prices were low. Assuming that the Ara Community Bank was to be operational by the end of 1994, women requiring small-scale loans would be in a position to withhold their products until a period of scarcity drove the prices upward. To eliminate the unfavourable market prices provided by traders coming from outside Ara to purchase palm products and *gari*, the women have discussed a variety of options that may address this marketing situation in the near future. Since Ara citizens are resident in numerous towns and cities in Nigeria, as well as the large contingent in the Ivory Coast, the women are considering working through the 19 branches of the Ara Descendants Union to establish a marketing information network. This would allow the Ara women to keep current with shifting market prices in the major urban centres located within a day's drive of Ara, such as Lagos, Ibadan, Ogbomosho, Ilorin, Ife and Oshogbo. Many of the Ara citizens living in these urban centres are involved in commercial enterprises. As the production of palm products and *gari* increases in Ara, these citizens of Ara could provide a steady and reliable marketing network for the Ara-based women. This would help to improve the incomes of Ara citizens within Ara and those living outside of Ara, providing the migrants with more means to invest in Ara development projects.

It was decided in summer 1993 by the Ara Branch of the Better Life for Rural Women programme that the new enterprise would require a separate organization. This was established and registered with the Egbedore local government as the Ara Women's Co-operative Food Products Enterprise. Established as a co-operative, any woman of Ara could join for a N250 share contribution. This was later raised to N500. Most of the share contributions were used to assist in the expenses of construction of the facility to house the machines and engines. It was decided that officers would serve one-year terms. Women who belong to the co-operative will have access to the machines at a nominal fee. All of the women are expected to learn how to operate and maintain the machines themselves. During any slack periods when members are not using the machines, non-members can pay a higher fee to use them. An account has been established at the bank. Fees will be retained in the bank account to cover operating expenses such as the purchase of petroleum and diesel and the payment of utility bills, maintenance costs such as the regular maintenance and repair of the machines and engines, and a capital fund for possible future expansion of operations that would require the purchase of additional machines.

A new agricultural issue for the community is the perceived possible short supply of palm nuts given the far greater production capacity due to the machines. The lands of the Alara are covered with oil palm trees, but many of them are now very old and very tall. Numerous young men in Ara are trained to climb these trees and cut down the clumps of palm nuts. This is dangerous and tedious work. Moreover, the women must pay these men for their labour. Participants in the 1991 workshop were aware that there were new hybrid varieties of oil palm that grew to a maximum height of about six feet, yet produced up to three times the number of clumps of palm nuts. It was decided to identify sources of these hybrid seedlings. This has been done and several hundred have now been acquired. Perhaps challenging tradition, some of the women have acquired seedlings and are nursing them on their own farms. The women's co-operative has also discussed the possibility of acquiring land through the Alara to establish their own oil palm plantation. How these initiatives evolve is yet to be determined.

These innovative ventures related to the management conditions for the processing of agricultural products have the potential greatly to improve the income and quality of life of women as well as households in general in Ara. The experiments with social organization and management of new enterprises have, however, not been without costs. The time and energy involved in raising funds have been considerable. After the first shipment of machines and engines arrived in Ara, some men within Ara told members of the women's co-operative that development was 'men's work' and that they would run the machines for the women. This intention resulted in a number of serious confrontations between these men and the women's co-operative that were finally resolved by the *Alara* to the satisfaction of the women – and the community in general.

Conclusions

There are several key lessons for development professionals based on the case-study of Ara. Although development agencies have supported management and development planning training projects in numerous countries, very little attention has been directed to improving the capacity of community-level development associations, which tend to remain invisible to many outsiders. Community leaders and citizens are very open and eager about exploring ways to improve their organizations for more effective development. It was discovered that a full range of management and development planning terms in English also exist in Yoruba, indicating that these concepts have been present in communities such as Ara for many generations. It was also discovered that illiterate and semi-literate workshop participants have no difficulty in handling experiential approaches to training. Particularly important to development agencies is the fact that citizens of a community can have a comprehensive understanding of both strengths and weaknesses within their own organizations and can be exceptionally open to trying new management and planning mechanisms that have proven successful in other communities. A development opportunity such as the 1991 workshop and the follow-up annual evaluative meetings have provided a new dynamism to the existing development organizations within Ara, spurring improved levels of support for community development projects by citizens within the community as well as the many Ara migrants living elsewhere. Citizens such as Mary Warren, living in the USA, have excited the interest and support for these community efforts by friends in the USA, Canada, and Japan, resulting in several of them visiting Ara personally. The fourth-grade classes in an Iowa elementary school even raised money that supported the extension of water lines to both the Ara Community Day Care and the Ara Community Library. Many of these fourth-graders are now pen pals with their Ara counterparts, something that will extend the horizons of the students in both Ara and Iowa, perhaps leading to improved international and cultural understanding.

Indigenous agricultural experimentation in a community is based on the array of perceived problems and opportunities expressed by different interest groups. As the case-study of Ara indicates, experimentation that impacts on the agricultural sector extends well beyond production agriculture to include socio-economic and technological conditions that limit both production agriculture and the processing of agricultural produce. Working with and through local-level indigenous organizations can provide opportunities for citizens of the communities (as well as outsiders) to understand better ways of strengthening these organizations while carrying out development projects designed to improve conditions identified by citizens themselves.

15. Chinese farmers' initiatives in technology development and dissemination: a case of a farmer association for rural technology development

LI XIAOYUN, LI OU and LI ZHAOHU

Abstract

IN THIS PAPER, the initiatives of Chinese farmers in technology development and dissemination will be illustrated by means of a case-study on a farmer association for rural technology development. Against the background of the agricultural and rural reforms in China, the reasons why farmers conduct their own on-farm research and disseminate the technologies they develop, will be examined. The main features and innovations in the process and the mechanisms of the operation will be introduced. Some conclusions are drawn and recommendations made for researchers of indigenous knowledge, development practitioners and policy-makers.

Introduction

In the 1980s, farmer associations for agricultural technology development began to expand very rapidly in rural China. Two reasons contributed to this development. First, after the commune-brigade system was dismantled at the beginning of the 1980s, the agricultural extension system network, consisting of county, commune, brigade and production teams, did not work any more. The recipients of technology shifted from the leaders of the brigades and teams to the millions of individual farm households, and at the same time funds and personnel for adaptive research and technology development at the field level were reduced.

Second, due to the implementation of the farm household responsibility system, farmers' production objectives to a certain degree contradicted the objectives of the government. Farmers wanted to increase their income through shifting their resources into enterprises with high value products, while the government still insisted on increasing the yield of grain crops to maintain China's food security. This led to a gap between the demand for relevant and appropriate technologies by farmers and the types of technologies supplied by government.

On the other hand, the rural reforms in China provided farmers with many opportunities to increase their income. The rapid increase in income levels stimulated the expansion and diversification of demand for agricultural products. The international market also created more demand for agricultural products. So many innovative farmers and farmer organizations emerged to develop appropriate technologies for farmers to take advantage of these opportunities. A lot of on-farm experiments in cash crops such as cotton, vegetable and fruit and animal production were carried out by farmers and farmer organizations.

In this paper, a case-study of a farmer association for rural technological development is used to illustrate how farmers determine their objectives, screen the scientific research, manage on-farm research, and evaluate and disseminate the research findings. Initiatives of farmers in technology development and dissemination have implications for research on indigenous knowledge.

The background and establishment of the Guoxin Association

The Guoxin Association is named after Mr. Lu Guoxin, a farmer and the director of the Association. The Association for Rural Technology Development was established in 1984. At the beginning it consisted of 12 farm households in Lucun Village, Hejian County, Hebei Province. Now it has about 2500 households from 79 villages in Hejian and three other neighbouring counties, covering about 1000 ha of land (the average farm size is about 0.6 to 0.8 ha in these areas).

The reason for its establishment was that the farmers wanted to retain their advantage in cotton and cotton seed production when the market appeared to discriminate against them. After Chinese farmers became independent managers of contracted land at the beginning of the 1980s, they quickly developed grain and cash crops – cotton in this case – which increased their income. By 1984, the national yields of grain crops and cotton had reached their peak and a surplus of supply appeared, due to the constraints of storage, processing and demands. So the government made its price policies and quotas more favourable to farmers' production, especially of cotton. The cotton stations stopped purchasing a cotton variety which they call Shandong Cotton No. 1, which had brought farmers a lot of income with its high yield, due to surplus supply, as well as the low quality of the variety. The seed production team led by Lu Guoxin in Lucun Village had to be dismantled, although the seed production of Shandong Cotton No. 1 had made profits. In order to obtain new varieties and technologies directly from university researchers and institutes and to obtain economies of scale in cotton ginning and seed supply, the Guoxin Association was established.

The characteristics and innovations of the Association in technology development and dissemination

The objectives of the Farmers' Guoxin Association are addressing farmers' needs and problems through on-farm research. The Association initiated a system combining the experience of researchers, technicians and farmers together for the design and implementation of on-farm experimentation. Because technology development and dissemination were closely related to the interests of the Association members, the efficiency and effectiveness was much higher than in government organizations.

Emphasizing the farmers' needs and problems

To maintain income levels or find new income opportunities, the Association selected cotton seed production in the first years and high added-value products such as fruit, vegetable and animals in later years as topics of their on-farm research.

Case 1 In the spring of 1985, the first growing season after the Association was established, the farmers identified two problems for their on-farm experiments in cotton production.

- *Variety*: The Agricultural Bureau in Hejian County determined that variety No. 8 of Hebei Cotton could be disseminated that year. But the farmers of

Guoxin Association had experienced the long growing period of this variety, its excessive post-frost flowers which decreased the quality, and its poor growth in rain-fed, infertile and saline-alkaline land.

- *Bad performance in the seedling stage*: in Hebei province, as well as in North China, irrigated and fertile land was always used for wheat production and bad land for cash crops such as cotton in the 1980s. It reflected the farmers' choice – food security as priority. However, the germination of cotton was slow and late, and pests and diseases were severe in the seedling stage on saline-alkaline land.

To solve the problems, the farmers screened the available varieties and technologies through adaptive research. Several new cotton varieties bred by the Chinese Institute for Cotton Research were selected. Plastic film was used to cover the land to increase the temperature and humidity and a coating agent was used to coat the seeds to prevent pests and disease and alleviate the nutrition deficit in the soil.

After the success in the first year, the farmers wanted to reduce the high level and arduous nature of labour involved in pruning cotton. Chemical control of cotton growth, developed by researchers of the Beijing Agricultural University, was selected to solve the problem. Recently, the technology was also combined with short growth-period varieties and high density planting to develop a winter wheat/summer cotton intercropping pattern, which brought more income to the Association members and has been disseminated extensively in North China.

Several years later, the comparative advantage of crop production decreased as a result of price increases in the inputs such as fertilizers and pesticides. A demand for adjustment of the cropping and farming structures emerged. The Association started to do research on cropping patterns and other high added-value enterprises. The technologies of cotton–fruit intercropping and cotton–vegetable multi-cropping were developed. Three other branch associations for fruit, vegetable and pig production were established.

Co-operation between researchers, technicians and farmers in on-farm research

The Association invited 25 experts from universities and institutes as consultants. Two government extensionists were hired as technicians. Fourteen innovative farmers, including Mr. Lu Guoxin, also work as technicians full-time or part-time. All of them have official titles such as senior agronomist, agronomist, assistant agronomist and technician.

The farmer technicians determine the topics of research and design the experiments before the growing season. Then the plans and designs are evaluated and revised by the technicians and experts together to apply the theories developed.

On-farm research is carried out in the fields of farmers who are contacted by the Association – fields always selected at convenient locations for other farmers to be able to visit easily and make their own evaluation. The contact farmers manage the fields according to the measures designed. Technicians share the responsibility for monitoring and recording the experiments, supervising the seed purity and advising the Association members, as well as other farmers among the 79 villages. They are also responsible for organizing training courses and field demonstrations.

Appropriate technologies have not only developed through adaptive research on the scientists' findings, but also from farmers' indigenous knowledge.

Case 2 Sex-pheremone agents were developed for pest control, but the effectiveness was not very satisfactory. The area affected was limited and the female pests were not affected. In 1992 a member farmer found by chance a lot of bollworm moths gathered on the flowers of the carrot and celery plants, sucking the honey. The moths were both male and female. After receiving this information, the Association decided to fund experiments on the effectiveness of these plants for trapping pests. On-farm research and demonstrations were carried out at the same time. An 'inducing plot' technology was developed and proved more effective than the commercial pheremone agents. A famous Chinese professor of plant protection gave it a very high rating.

High efficiency in technology development and dissemination

The technicians of Guoxin Association serve the interests of the members as well as their own interests, and have therefore proved more efficient than the government extensionists.

Case 3 For integrated pest management (IPM) the technicians continually do experiments on the effectiveness of pesticides and monitor the density of insects. In the laboratory the technicians feed different kinds of insects at different developmental stages, combine different insecticides, and adjust the composition of ingredients to find the best ones for pest control. According to the experimental results, the Association then purchases a sufficient amount of pesticides for preparation. In the fields, the technicians monitor the pest density. If the density exceeds the standard for IPM, the information will be promptly disseminated. The next day, pesticides will be sent to the Association members, and compounded and spread under the guidance of the technicians. The Agricultural Bureau of Hejian County also uses the information provided by the Association to guide pest control in the whole county. But that operation is much slower because of the many intermediate links (county – townships – villages – households) and the low motivation of the government extensionists.

Through the co-operation between researchers and the Association, four suitable improved cotton varieties were selected from 73 varieties, and 16 appropriate technologies were developed through adaptive research. It takes only two to three years from adaptive research and dissemination of findings about appropriate technology to the large-scale use of the technology by farmers. This is much faster than in most other cases of technology transfer in China.

Innovative mechanisms of technology development and dissemination

Some innovative mechanisms were developed by the Guoxin Association. The Association pays the researchers for their new varieties and research findings and funds their experiments on the farmers' fields. It also uses economic interests to motivate farmers to participate in the experiments and apply the technologies. All the funds for experiments and technology dissemination were accumulated through the services of the Association.

Purchase of research findings for technology development

In the 1980s, the Chinese government requested researchers to try to solve farmers' practical problems and create some model areas for dissemination of their research findings. However, the funds allocated were never enough for researchers to do both adaptive research and technology dissemination, so the Guoxin Association grasped the opportunity.

Case 4 The Chinese Institute for Cotton Research supplies seeds of new-bred varieties to clients on condition that the Institute shares 5 to 10 per cent of the net profit the clients gain in their seed production. The state seed companies always break the promise because the present managers seldom think about the long-term development of the companies. As an NGO, the Guoxin Association can only accumulate funds and exist with participation of the members themselves through using advanced technology, so they appreciate the value of science and scientists and many cotton breeders have become good friends of the farmers in the Association. Some of them bought the newest varieties for adaptive research. The Association uses this advantage to earn more money in seed production.

Case 5 The cotton bollworm is the worst enemy of cotton farmers in North China. Sometimes it destroys 30 to 70 per cent of the production. In 1992, a bad infestation of this insect plague occurred, so the Association bought a prescription of compound insecticides developed by a professor from Nankai University (located in Tianjin) for 15 000 RMB (1 US$ = 8.7 RMB, 1994). The Association paid him another 10 000 RMB for further experiments on bollworm and also paid him for new prescriptions.

In the same season, the Association also applied for assistance from research institutes for the production and supply of pesticides, through several national or provincial newspapers. The Association promised that if anyone could help the farmers control the cotton bollworm, he or she would be rewarded. Consequently, 27 institutes sent more than 30 kinds of new pesticides. Several effective ones were screened through eight experiments, and then supplied to the farmer members and 15 000 mu of cotton production was protected (1 Chinese mu = 0.067 ha).

Motivation for experiments and use of technology

Chinese farmers, like farmers all over the world, are very realistic. They believe in practical experience and pursue what they can gain. So the Association tried to find the means to motivate the farmers to do experiments and use the developed technologies.

For experiments, the Association not only covers all the costs, but also subsidizes the farmers up to a value equal to the highest profit the farm can produce. If there is a loss due to wrong design or guidance, the Association will repay the farmer. Therefore the farmers like to conduct experiments in their fields, because their risk is minimal and they can receive the benefits of service and knowledge.

Dissemination of technology is not the objective, but rather the means to increase farmers' income. The Association tried to reach that goal. In the 1980s the cotton purchasing and inputs supply were exclusively implemented by the state supply and marketing agencies. As a technical organization for farmers supported by government, the Association has succeeded in getting the rights of purchase and supply among its members. The farmer members can therefore get a higher price for the cotton they produce than the market price because the quality is high and the seed worth more. In a bad year for the market price of cotton, the Association gives guaranteed prices to the farmers because the Association has made direct contracts with textile mills and received favourable prices.

Regarding inputs, the Association has supplied 90 per cent of pesticides and 40 per cent of fertilizers that the members used in recent years. For new technical inputs, such as seed coating agents and chemical controllers for cotton growth, 100 per cent of them were supplied by the Association. The Association gets the source of inputs at wholesale prices or state prices. So the price within the Association for pesticides can be cheaper by 10 to 15 per cent. For fertilizers the farmer members can get them at state prices, which are 20 to 40 per cent lower than market prices. In the 1980s the inputs had two prices in China. Farmers could buy inputs at the lower one, the state price, for the quota of grain and cotton they produced and sold to the state.

From one mu of cotton production, the farmer members obtained twice the profit that non-member neighbours gained on average over the last 10 years, because the yield per mu was higher by one-third and the labour used on the field (for pest control and pruning cotton) less than half their neighbours', due to the application of technology.

Build-up of funds from commercial activities to provide better services

The strategy of the Guoxin Association is to accumulate funds through use of improved technology and the advantages of improved economies of scale. The funds are used for experiments and technology dissemination.

The activities of the Association require a lot of funding. For example, the Association purchases inputs at wholesale prices for later use by the members for about 300 000 RMB annually. So the Association has set up a factory to gin cotton and supply seeds of improved varieties. Contracts were signed between the members or village group leaders and the Association. The farmers should guarantee the seed purity of the cotton growing in their fields and sell all the cotton they produce to the Association. Ten per cent profit on seed production was gained by the Association. The Association has also financed a building, half of which is used for training and other services, and the remaining half for accommodation as a guesthouse. It has also set up its own farm for experiment and demonstration. In 1991, the profit from the commercial activities earned 250 000 RMB in total for the activities of the Association.

In 1989, the Association developed a stock-sharing co-operative arrangement. Thirty-two farmer members put 485 000 RMB in total as the capital for the new building. According to the constitution, the stock-holders can share 20 per cent of the profit made by the guesthouse. But for the sake of the Association, they voluntarily gave up the shares until the end of 1995 for reinvestment.

Conclusions and recommendations

The case of the Guoxin Association shows us a completely different picture of technology development and dissemination from that of government research and extension organizations in China. In summary, it illustrates that the farmers organized themselves for their own sake, initially introduced the research findings of the professors and researchers, adapted them to farmers' needs, resource base and socio-economic environment, and disseminated the appropriate technologies or farming systems that they developed to the hundreds of thousands of farm households in areas where there were no funds allocated or personnel assigned by the state. Some implications and conclusions can be drawn for researchers of indigenous knowledge, rural development practitioners and policy-makers.

- The fact that the Gouxin Association has done so much on-farm research and developed so many appropriate technologies, mainly in one commodity, in the last 10 years indicates that there is an urgent need to bridge the gap between the research findings and the farmers' needs, circumstances and practices for agricultural and rural development in China. However, most of the policy-makers for research management ignored the existence of the gap and the importance of on-farm research, especially with farmers' participation. They considered that to include adaptive research and to establish a certain number of model extension areas in a research project is enough for technology development. They confused the research findings with appropriate technology.
- One of the most effective approaches to bridge the gap is to develop farmers' organizations. An efficient mechanism for introducing research findings, doing on-farm research, disseminating technology and accumulating funds for all of these activities should be developed within the organization. The development of such farmers' organizations can complement the services from government research and extension organizations and help farmers transfer new knowledge and technologies into practice. However, striking contrasts can now be found in China. Many important research projects at the national level can only be carried out in some places if the local community benefits from the funds, otherwise all the impact may vanish in the research area once the project is finished.
- Based on a reconsideration of the strategies for scientific and technological research management and the roles of the farmers and farmer organizations in technology development and dissemination, policy-makers should conceive of a completely new system of knowledge and technology, give farmers an appropriate position in the system, and design a favourable policy for them to acheive their full potential.

Notes

Chapter 2

1. The authors are, respectively, Anthropologist, Economics Program, International Maize and Wheat Improvement Center (CIMMYT); and Ecologist, Colegio de Postgraduados, Montecillos, Mexico. The views represented in this paper are the authors' and do not necessarily reflect policies of their respective institutions.
2. An *ejido* is both a specific territory and an association of producers (*ejidatorios*) with rights to use land owned by the *ejido* members. While land ownership is collective, in most cases *ejidatarios* work specific parcels of land independently.
3. According to family members, the late Don Faustino Hernandez, a Nahua farmer from Mecayapan, and later Mirador Saltillo in the same municipality, brought velvetbean seed to Pozo Blanco in the late 1940s from his father's homeland in Anancajucan, Tabasco, a region where velvetbean is also used (cf. Granados, 1989). Various other early users of velvetbean in the region claim to have 'invented' uses of the plant independently.
4. After the first cycle, various farmers became involved in a community-wide land tenure dispute which forced them to move to new maize fields (*milpas*). Some of these farmers started the trials over again on their new fields, but others did not. Two new farmers relocated on to fields with established velvetbean trials chose to continue them in collaboration with the researchers. The velvetbean crop developed very poorly on some fields, prompting these farmers to abandon the trial. One farmer died of a sudden illness; several others temporarily abandoned farming to take up non-farm employment.
5. Two participating farmers (Reynaldo Pantaleon and Gustavo Antonio) were hired part-time to locate farmers' fields, notify farmers of meetings, and help collect data. The authors gratefully acknowledge their contributions.
6. In two of these sites an experiment was sown the second year to determine why velvetbean had not developed. In one field, poor growth was overcome with weeding while in the other field velvetbean growth was enhanced by small amounts of fertilizer and inoculation of velvetbean seed with suitable Rhyzobia.
7. The description of farmers' perspectives on shifting cultivation is based on Chevalier and Buckles (1995).

Chapter 4

1. According to Hillel (1992) this behavioural adaptation and inventiveness has contributed much to the success of the human species, for the rapid venture into very diverse climates and landscapes did not allow an entirely genetic or physical adaptation.
2. Fieldwork took place within the framework of the Water Spreading Research Kassala (WARK) programme of the Sudanese National Council for Research and was conducted under supervision of J. A. van Dijk.
3. Namely farming areas of the villages Um Safaree, Hafarat, and Ilat Ayot (see Figure 4.1).
4. For a more extensive overview of the historical developments in the area see Niemeijer (1993).
5. The first evidence of sorghum cultivation dates back to 1500 BC (Van Dijk, 1995).
6. Classified according to the revised FAO classification of 1988.
7. According to Van Dijk (1995) the first evidence of *teras* cultivation in the Border Area dates from reports and aerial photographs from the early 1950s. It is likely, however, that the actual use of the *teras* technique predates this evidence by a few years.
8. In earlier years the proportion must have been even larger, for in many areas large-scale colonial irrigation schemes had replaced *teras* cultivation by the 1940s.

9. There is no long-term yield data of statistical quality available for the marginal Border Area that would allow the calculation of yield frequencies over time.
10. Because of cultural and religious reasons arable farming is almost exclusively in the male domain and, though women occasionally assist during the harvest, there are no female farmers. For that reason gender-specific terms are used throughout this paper.
11. Östberg (1995) documented similar, albeit smaller, field-level water management techniques among some Burunge farmers of central Tanzania.
12. In terms of household wealth other activities are much more important – notably, livestock-related activities that contribute an average 75 per cent to household wealth (Van Dijk and Mohamed Hassan Ahmed, 1993).

Chapter 6

1. 'Livelihood for the People' (*Kauyagan ho Kahilawon*).
2. For the present analysis, we consider, on the basis of close observation of the material in both farms and in an experimental station genebank, and via detailed discussions with farmers, that differently named cultivars represent different genotypes and like-named cultivars are clones. This still merits confirmation either through further characterization, or using molecular techniques.

Chapter 8

1. A report on the Maheshwaram Project laments: ' ... the whole programme was viewed as a government activity and no farmer has participated in the execution works.' Pilot Project for Watershed Development in Rainfed Areas: Maheshwaram. Administrative Staff College of India, Hyderabad. p. 54.
2. Around 30 representatives of voluntary agencies and others met at a People's Science Forum in Bangalore, June 1989. Report available from the Deccan Development Society, A-6, Meera Apts., Basheerbagh, Hyderabad 500 029, A.P., India.
3. We refer to the discourses on Orientalism (Edward Said) and Subaltern studies (Ranajit Guha) that are giving a new air of confidence to work rooted in Third World perspectives. Compare also the Feminist discourse on parallel issues – e.g. S. Harding: *The Science Question in Feminism*, Open University Press, U.K, 1986.
4. Most important is Dharampal's *Indian Science and Technology in the 18th Century*, Academy of Gandhian Studies, Hyderabad, 1983.
5. See, for example, K. P. Kannan's 'Secularism & People's Science Movement in India', *Economic & Political Weekly*, Feb 10, 1990.
6. John Perkins: *Pests, Experts, and Insecticides*, John Wiley, New York, 1982. An excellent book that gives a chatty but scientific description of the struggles within the life sciences to bring a more rational and ecological approach to pest management. Of particular note is the work of pathbreaking scientists who introduced plants to attract pests (!), which in turn would attract controlling populations of predators.
7. Refer to 'Participatory Research with Women Farmers', a scientific film produced by Development Perspectives, for Dr. Michel Pimbert, International Crop Research Institute for the Semi-Arid Tropics, Patancheru, A.P., India.
8. Jan Seva Mandal recorded great success in mobilizing local children to collect and destroy the clearly visible yellow egg masses of the pest.
9. Pat Mooney was given the Right Livelihood Award, also known as the Alternative Nobel Prize, in 1986 for his work in alerting the Third World to the danger of germplasm concentration in the hands of multinationals and Western research institutes.
10. Many power groups struggle for ascendency in a developing country, particularly in a rural situation. We may broadly identify four tendencies: one is led by the 'modernists', who include scientists, government officials and businessmen, who believe in the liberating forces of science and modern organization. Another is composed of Marxists who also believe in science but want people's control. A third force is that of 'traditionalists' who decry modernisms, but otherwise join hands with the powerful. A fourth is made up of 'romantics' who walk part of the way with the others, but support none. Under these conditions, the search for 'purity,' with its special Indian meanings, blocks attempts at practical work.

11. Light traps used by entomologists employ fuming chemicals to kill insects. Water can be used to trap the moths, with a soapy film to lessen surface tension. Design, size and safety measures need to be worked out for rugged rural conditions. Also, power failure constantly interrupts efficiency, when most needed. Can gas lights be used just as effectively? Further, it has been noted that moths are more attracted to the blue frequencies of light, and that the normal florescent tube-lights, used as stand-bys in India during power break-downs, could very easily serve the purpose. The lights could also be used for night schools, village clinics, meetings, and hence be more affordable to the village communities.
12. M. A. Qayum and N. K. Sanghi (eds) *Red Hairy Caterpillar Management through Group Action and Non-Pesticidal Methods*, ASW and Oxfam (India) Trust.
13. See M. K. Mukundan's excellent study on traditional village tank systems. Patriotic and People-Based Science & Technology Foundation (PPST), P.O. Box 2085, Adyar, Madras, India.
14. Personal communication from Dr. T. Hanumantha Rao, Ex-Director General, Water and Land Management Research Institute, Hyderabad.
15. Dr. N. K. Sanghi, Central Research Institute for Dryland Agriculture (CRIDA), Hyderabad, India, reports that some indigenous peoples in Adilabad District, A.P., are innovating in field bunding practices. Women *sangams* collectively leasing in land in Medak District also innovate for soil control – see Deccan Development Society's reports.
16. ICRISAT: *Annual Report 1986*.
17. Irfan Habib: *An Atlas of the Mughal Empire*, Oxford University Press, Oxford 1982.
18. Permaculture, a concept for designing organic farming in harmony with nature, was developed by Bill Mollison of Australia, who also won the Alternative Nobel Prize. Permaculture has struck firm roots among Indian NGOs. A Permaculture Association has been formed sharing an office with the Deccan Development Society, at Hyderabad.
19. J. D. Bernal, science historian, used the term 'people-based science' but perhaps not too appropriately considering the divergence of interests between common people and the scientific community, especially physicists, living off defence research contracts.

Chapter 13

I would like to dedicate this chapter to the swamp farmers of Maputaland, especially David Mathenjwa, Joshua Mathenjwa and Charles Mlambo from whom I learnt about farming in Maputaland.

1. 'Umnotho wethu amadobo' translated from Zulu means 'the swamp/swampy areas/ swamp forests are our fertility/wealth'. The old man who made this statement was referring to the swamp forests which he had utilized in the past and which had been used by his father before him.
2. KwaZulu was the Bantustan administered in a form of quasi-independence by the KwaZulu Government and the ethnic base for the Inkatha Freedom Party, both headed by Chief Buthelezi. It is a fragmented territory spread throughout Natal and includes Maputaland. Zululand is the area north of the Tugela River and was traditionally the area of the Zulu kingdom. Presently Zululand is composed of parts of KwaZulu and parts of Natal. The new region incorporating Natal and KwaZulu is now known as KwaZulu-Natal.
3. I refer to people as Thonga and Zulu to emphasize cultural and linguistic differences and not to perpetuate ethnicity.
4. According to Bryant (1949), uNyawoti (pearl millet) was formerly widely grown by the coastal Zulus having been replaced during the last century by the more prolific amaBele (sorghum) which in turn has been largely replaced by umBila (maize). Pearl millet is almost unknown in Zululand now. But he is in all likelihood incorrect, for both millets and sorghums would have formed part of the farmer's repertoire in both Zululand and Maputaland.

References

Introduction

Adams, William M. and L. Jan Slikkerveer, eds. (1997) *Indigenous Knowledge and Change in African Agriculture.* Studies in Technology and Social Change, No. 26. Ames: CIKARD, Iowa State University.

Adegboye, Rufus O. and J. A. Akinwumi (1990) 'Cassava Processing Innovations in Nigeria.' In Matthew S. Gamser, Helen Appleton, and Nicola Carter, eds. *Tinker, Tiller, Technical Change.* London: Intermediate Technology Publications.

Amanor, Kojo S. (1989) 340 abstracts on farmer participatory research. *Agricultural Administration (Research and Extension) Network Paper 5.* London: Overseas Development Institute.

—— (1991) 'Managing the Fallow: Weeding Technology and Environmental Knowledge in the Krobo District of Ghana.' *Agriculture and Human Values* 8 (1+2): 5–13.

Ashby, Jacqueline A., Teresa Gracia, Maria del Pilar Guerrero, Carlos Arturo Quiros, Jose Ignacio Roa and Jorge Alonso Beltran (1996) 'Innovation in the Organization of Participatory Plant Breeding.' In Pablo Eyzaguirre and Masa Iwanaga, eds., *Participatory Plant Breeding.* Rome: IPGRI.

Ashby, Jacqueline A., Carlos A. Quiros and Yolanda M. Rivers (1989) 'Farmer Participation in Technology Development: Work with Crop Varieties.' In Robert Chambers, Arnold Pacey, and Lori Ann Thrupp, eds. *Farmer First: Farmer Innovation and Agricultural Research.* London: Intermediate Technology Publications.

Bentley, J. W. (1994) 'Facts, fantasies, and failures of farmer participatory research.' *Agriculture and Human Values.* Spring-Summer: 140–150.

Berg, Trygve (1996) 'The Compatibility of Grassroots Breeding and Modern Farming.' In Pablo Eyzaguirre and Masa Iwanaga, eds. *Participatory Plant Breeding*. Rome: IPGRI.

Biggs, Stephen D. (1980) 'Informal R&D.' *Ceres*: 13:4:23–26.

Biggs, Stephen D. and Edward Clay (1981) 'Sources of Innovation in Agricultural Technology.' *World Development* 9: 321–336.

Box, Louk (1989) 'Virgilio's Theorem: A Method for Adaptive Agricultural Research.' In Robert Chambers, Arnold Pacey, and Lori Ann Thrupp, eds. *Farmer First: Farmer Innovation and Agricultural Research.* London: Intermediate Technology Publications.

Brokensha, David W., D. M. Warren, and Oswald Werner, eds. (1980) *Indigenous Knowledge Systems and Development.* Lanham, MD: University Press of America.

Brouwers, J. H. A. M. (1993) *Rural People's Response to Soil Fertility Decline: The Adja Case (Benin).* Wageningen Agricultural University Paper No. 93–4. Wageningen: Agricultural University.

Brush, Stephen B. and Doreen Stabinsky, eds. (1996) *Valuing Local Knowledge: Indigenous People and Intellectual Property Rights.* Washington, D.C.: Island Press.

Bunch, Roland (1985) *Two Ears of Corn: A Guide to People-centered Agricultural Improvement.* 2nd ed. Oklahoma City: World Neighbors.

—— (1989) 'Encouraging Farmers' Experiments.' In Robert Chambers, Arnold Pacey, and Lori Ann Thrupp, eds. *Farmer First: Farmer Innovation and Agricultural Research,* London: Intermediate Technology Publications.

—— (1990) *Low Input Soil Restoration in Honduras: Cantarranas Farmer-to-Farmer Extension Programme.* Gatekeepers Series, No. 23. London: International Institute for Environment and Development.

Bunders, Joske, Bertus Haverkort and Wim Hiemstra, eds. (1997) *Biotechnology: Building on Farmers' Knowledge.* Basingstoke, UK: Macmillan Education.

Center for Traditional Knowledge (1997) *Guidelines for Environmental Assessments and Traditional Knowledge.* Ottawa: Centre for Traditional Knowledge. Unpublished MS.

Chambers, Robert (1983) *Rural Development: Putting the Last First.* London: Longman.

—— (1997) *Whose Reality Counts? Putting the First Last.* London: Intermediate Technology Publications.

Chambers, Robert, Arnold Pacey, and Lori Ann Thrupp, eds. (1989) *Farmer First: Farmer Innovation and Agricultural Research.* London: Intermediate Technology Publications.
Conklin, Harold C. (1957) *Hanunoo Agriculture: A Report on an Integral System of Shifting Cultivation in the Philippines.* Forestry Development Paper 12. Rome: FAO.
de Boef, Walter, Kojo Amanor and Kate Wellard, with Anthony Bebbington (1993) *Cultivating Knowledge: Genetic Diversity, Farmer Experimentation and Crop Research.* London: Intermediate Technology Publications.
den Biggelaar, Christoffel (1991) 'Farming Systems Development: Synthesizing Indigenous and Scientific Knowledge Systems.' *Agriculture and Human Values* 8 (1+2): 25–36.
den Biggelaar, Christoffel and N. Hart (1996) *Farmer Experimentation and Innovation: A Case Study of Knowledge Generation Processes in Agroforestry Systems in Rwanda.* Rome: FAO.
Denevan, W. M. (1983) 'Adaptation, Variation and Cultural Geography.' *Professional Geographer* 35 (4): 399–406.
Dvorak, Karen Ann, ed. (1993) *Social Science Research for Agricultural Technology Development.* Wallingford: CAB International.
Evans-Pritchard, E. E. (1940) *The Nuer: A Description of the Modes of Livelihood and Political Institutions of a Nilotic People.* Oxford: Clarendon Press.
Eyzaguirre, Pablo and Masa Iwanaga, eds. (1996) *Participatory Plant Breeding.* Proceedings of a workshop on participatory plant breeding, 26–29 July 1995, Wageningen, the Netherlands. Rome: International Plant Genetic Resources Institute (IPGRI).
Farrington, J. and A. Martin (1987) *Farmer Participatory Research: A Review of Concepts and Practices.* Discussion Paper 19, Agricultural Administration (Research and Extension) Network. London: Overseas Development Institute.
Forde, C. Daryll (1949) *Habitat, Economy and Society: A Geographical Introduction to Ethnology.* New York: E. P. Dutton.
Gamser, Matthew S., Helen Appleton, and Nicola Carter, eds. (1990) *Tinker, Tiller, Technical Change.* London: Intermediate Technology Publications.
Haverkort, B., van der Kamp, J. and Ann Waters-Bayer, eds. (1991) *Joining Farmers' Experiments.* London: Intermediate Technology Publications.
Hiemstra, Wim with Coen Reijntjes and Erik van der Werf (1992) *Let Farmers Judge – Experiences in Assessing Agriculture Innovations.* London: Intermediate Technology Publications.
IIRR (1996) *Recording and Using Indigenous Knowledge: A Manual.* Silang, Cavite, Philippines: REPPIKA, International Institute of Rural Reconstruction.
Innis, Donald Q. (1997) *Intercropping and the Scientific Basis of Traditional Agriculture.* London: Intermediate Technology Publications.
Johnson, A. W. (1972) 'Individuality and Experimentation in Traditional Agriculture.' *Human Ecology* 1 (2): 149–159.
Joshi, A. and J. R. Witcombe (1996) 'Farmer Participatory Crop Improvement. II. Participatory Varietal Selection, A Case Study in India.' *Experimental Agriculture* 32: 461–477.
King, Franklin Hiram (1911) *Farmers of Forty Centuries, or, Permanent Agriculture in China, Korea and Japan.* Madison: Mrs. F. H. King.
Leach, Edmund R. (1968) *A Runaway World?* New York: Oxford University Press.
Leakey, Richard E. and L. Jan Slikkerveer, eds. (1991) *Origins and Development of Agriculture in East Africa: The Ethnosystems Approach to the Study of Early Food Production in Kenya.* Studies in Technology and Social Change, No. 19. Ames: CIKARD, Iowa State University.
Matlon, P. J. (1988) 'Technology evaluation: five case studies from west Africa.' In: Matlon, P., R. Cantrell, D. King, and M. Benoit-Cattin, eds. *Coming Full Circle: Farmers' Participation in the Development of Technology.* Ottawa: International Development Research Centre.
McCorkle, Constance M. (1994) *Farmer Innovation in Niger.* Studies in Technology and Social Change, No. 21. Ames, Iowa: Center for Indigenous Knowledge for Agriculture and Rural Development, Iowa State University.

McCorkle, Constance M. and Hernando Bazalar (1996) 'Field Trials in Ethnoveterinary R&D: Lessons from the Andes.' In Constance M. McCorkle, Evelyn Mathias, and Tjaart W. Schillhorn van Veen, eds. *Ethnoveterinary Research and Development*. London: Intermediate Technology Publications.

McCorkle, Constance M. and Gail McClure (1995) 'Farmer Know-how and Communication for Technology Transfer: CTTA in Niger.' In D. Michael Warren, L. Jan Slikkerveer and David Brokensha, eds. *The Cultural Dimension of Development: Indigenous Knowledge Systems*. London: Intermediate Technology Publications.

Moock, Joyce Lewinger and Robert E. Rhoades, eds. (1992) *Diversity, Farmer Knowledge, and Sustainability*. Ithaca: Cornell University Press.

National Research Council (1991) *Toward Sustainability: A Plan for Collaborative Research on Agriculture and Natural Resource Management*. Washington, D.C.: National Academy Press.

—— (1992) *Conserving Biodiversity: A Research Agenda for Development Agencies*. Washington, D.C.: National Academy Press.

Netting, Robert (1968) *Hill Farmers of Nigeria: Cultural Ecology of the Kofyar of the Jos Plateau*. Seattle: University of Washington Press.

Page, W. W. and P. Richards (1977) 'Agricultural pest control by community action: the case of the variegated grasshopper in southern Nigeria.' *African Environment* 2 & 3: 127–141.

Prain, Gordon and C. P. Bagalanon, eds. (1994) *Local Knowledge, Global Science and Plant Genetic Resources: Towards a Partnership*. Los Banos: UPWARD.

Pretty, Jules N. (1991) 'Farmers' Extension Practice and Technology Adaptation: Agricultural Revolution in 17th-19th Century Britain.' *Agriculture and Human Values* 8 (1+2): 132–148.

—— (1995) *Regenerating Agriculture: Policies and Practice for Sustainability and Self-Reliance*. London: Earthscan.

Rappaport, Roy A. (1968) *Pigs for the Ancestors: Ritual in the Ecology of a New Guinea People*. New Haven: Yale University Press.

Reijntjes, Coen, Bertus Haverkort and Ann Waters-Bayer (1992) *Farming for the Future: An Introduction to Low-External-Input and Sustainable Agriculture*. Leusden: ILEIA.

Rhoades, Robert E. (1987) *Farmers and Experimentation*. Discussion Paper 21, Agricultural Administration (Research and Extension). London: Overseas Development Institute.

Rhoades, Robert E. and Anthony Bebbington (1995) 'Farmers Who Experiment: An Untapped Resource for Agricultural Research and Development.' In D. Michael Warren, L. Jan Slikkerveer and David Brokensha, eds. *The Cultural Dimension of Development: Indigenous Knowledge Systems*. London: Intermediate Technology Publications.

Rhoades, Robert E. and R. Booth (1982) 'Farmer-Back-to-Farmer: A Model for Generating Acceptable Agricultural Technology.' *Agricultural Administration* 11: 127–137.

Richards, Paul (1985) *Indigenous Agricultural Revolution: Ecology and Food Production in West Africa*. Boulder: Westview Press.

—— (1986) *Coping with Hunger: Hazard and Experiment in an African Rice-farming System*. London: Allen and Unwin.

Sauer, C. O. (1969) *Seeds, Spades, Hearths and Herds: The Domestication of Animals and Foodstuffs*. Cambridge: Massachusetts Institute of Technology Press.

Scarborough, Vanessa, Scott Killough, Debra Johnson and John Farrington (1997) *Farmer-led Extension: Concepts and Practices*. London: Intermediate Technology Publications.

Scoones, Ian and John Thompson, eds. (1994) *Beyond Farmer First: Rural People's Knowledge, Agricultural Research and Extension Practice*. London: Intermediate Technology Publications.

Selener, Daniel, Jacqueline Chenier, Raul Zelaya *et al.* (1997) *Farmer-to-Farmer Extension: Lessons from the Field*. New York: International Institute of Rural Reconstruction.

Sillitoe, Paul (1998) 'The Development of Indigenous Knowledge: A New Applied Anthropology.' *Current Anthropology* 39 (2): 223–253.

Steward, Julian Haynes (1955) *Theory of Culture Change: The Methodology of Multilinear Evolution*. Urbana: University of Illinois Press.

Sthapit, B. R., K. D. Joshi and J. R. Witcombe (1996) 'Farmer Participatory Crop Improvement. III. Participatory Plant Breeding, A Case Study for Rice in Nepal.' *Experimental Agriculture* 32: 479–496.

Sumberg, James and Christine Okali (1997) *Farmers' Experiments: Creating Local Knowledge*. Boulder: Lynne Rienner Publishers.

Systemwide Programme on Participatory Research and Gender Analysis (1997) *A Global Programme on Participatory Research and Gender Analysis for Technology Development and Organisational Innovation*. AgREN Network Paper No. 72. London: Agricultural Research and Extension Network, UK Overseas Development Administration (ODA).

Thurston, H. David (1992) *Sustainable Practices for Plant Disease Management in Traditional Farming Systems*. Boulder: Westview Press.

—— (1997) *Slash/Mulch Systems: Sustainable Methods for Tropical Agriculture*. Boulder: Westview Press and London: Intermediate Technology Publications.

Thurston, H. David and Joanne M. Parker (1995) 'Raised Beds and Plant Disease Management.' In D. Michael Warren, L. Jan Slikkerveer and David Brokensha, eds. *The Cultural Dimension of Development: Indigenous Knowledge Systems*, London: Intermediate Technology Publications.

Thompson, S. (1973) *Pioneer Colonisation: A Cross Cultural View*. Addison-Wesley Module in Anthropology, No. 33. Reading, MA: Addison-Wesley Publishing Company.

Tripp, Robert. (1989) *Farmer participatory research in agricultural research: new directions or old problems?* Brighton: Institute of Development Studies Discussion Paper 256.

Van Veldhuizen, Laurens, Ann Waters-Bayer and Henk De Zeeuw (1998) *Developing Technology with Farmers: A Trainer's Guide for Participatory Learning*. London: Zed Books (in preparation).

Van Veldhuizen, Laurens, Ann Waters-Bayer, Ricardo Ramirez, Deb Johnson, and John Thompson, eds. (1997) *Farmers' Experimentation in Practice: Lessons from the Field*. London: Intermediate Technology Publications.

Warren, D. Michael (1989) 'Linking Scientific and Indigenous Agricultural Systems.' In J. Lin Compton, ed. *The Transformation of International Agricultural Research and Development*. Boulder: Lynne Rienner Publishers.

—— (1991) *Using Indigenous Knowledge in Agricultural Development*. World Bank Discussion Papers No. 127. Washington, D.C.: The World Bank.

—— (1994) 'Indigenous Agricultural Knowledge, Technology, and Social Change.' In Gregory McIsaac and William R. Edwards, eds. *Sustainable Agriculture in the American Midwest*. Urbana: University of Illinois Press.

—— (1996) 'The Role of Indigenous Knowledge and Biotechnology in Sustainable Agricultural Development.' In *Indigenous Knowledge and Biotechnology*. Ile-Ife, Nigeria: Indigenous Knowledge Study Group, Obafemi Awolowo University.

Warren, D. Michael, Remi Adedokun and Akintola Omolaoye (1996) 'Indigenous Organizations and Development: The Case of Ara, Nigeria.' In Peter Blunt and D. Michael Warren, eds. *Indigenous Organizations and Development*. London: Intermediate Technology Publications.

Warren, D. Michael and Jennifer Pinkston (1997) 'Indigenous African Resource Management of a Tropical Rainforest Ecosystem: A Case Study of the Yoruba of Ara, Nigeria.' In Fikret Berkes and Carl Folke, eds. *Linking Social and Ecological Systems*. Cambridge: Cambridge University Press.

Warren, D. Michael, L. Jan Slikkerveer, and David W. Brokensha, eds. (1995) *The Cultural Dimension of Development: Indigenous Knowledge Systems*. London: Intermediate Technology Publications.

Willcocks, Theo. J. and Francis N. Gichuki, eds. (1996) *'Conserve Water to Save Soil and the Environment'*: Proceedings of an East African Workshop on The Evaluation of Indigenous Water and Soil Conservation Technologies and the Participatory Devel-

opment and Implementation of an Innovative Research and Development Methodology for the Provision of Adoptable and Sustainable Improvements. SRI Report No. IDG/96/15. Bedford: Silsoe Research Institute.
Witcombe, John R. (1996) 'Participatory Approaches to Plant Breeding and Selection.' *Biotechnology and Development Monitor* No. 29: 2–6.
Witcombe, John R., A. Joshi, K. D. Joshi and B. R. Sthapit (1996) 'Farmer Participatory Crop Improvement. I. Varietal Selection and Breeding Methods and Their Impact on Biodiversity.' *Experimental Agriculture* 32: 445–460.

Chapter 1

Brokensha, D., D. Warren and O. Werner, (eds) (1980) *Indigenous Knowledge Systems and Development*, Lanham, Maryland: University Press of America.
Department of Food and Agricultural Marketing Services (DFAMS). (1990) *Agricultural statistics of Nepal*, Ministry of Agriculture, HMG, Nepal, Kathmandu.
Maurya, D. M., A. Botrell and J. Farrington. (1988) 'Improved livelihoods, genetic diversity, and farmer participation: a strategy and rice breeding in rainfed areas of India', *Experimental Agriculture* 24: 3: 311–320.
Osborn, T. (1990) Multi-institutional approaches to participatory technology development. London: Overseas Development Institute, *Agricultural Administration (Research and Extension) Network Paper 13.*
Prakash-Asante, K., S. Sandhu, and D. S. C. Spencer, (1984) Experiences with rice in West Africa. In: Matlon, P., R. Cantrell, D. King and Benoit-Cattin, (eds) *Coming Full Circle*. Ottawa: IDRC.
Rhoades, R., and R. Booth, (1982) Farmer-back-to-farmer: a model for generating acceptable agricultural technology. *Agricultural Administration 11*: 127–37.
Richards, P. (1985) *Indigenous Agricultural Revolution*. London: Century Hutchinson.
—— (1986) *Coping with Hunger: Hazard and Experiment in an African Rice-garming System*. London: Allen and Unwin.
—— (1987) 'Experimenting farmers and agricultural research'. Unpublished paper.
—— (1989) *The Spice of life? Rice varieties in Sierra Leone*. Dept. of Anthropology, University College London.
Thurston, H. D. (1990) Plant disease management practices of indigenous farmers. *Plant Disease* 74: 2: 96–102.
Warren, D. M. and Cashman, (1988) *Indigenous knowledge for sustainable agricultural development*. Gatekeeper SA 10: Int. Inst for Env & Dev.

Chapter 2

Arévalo, Ramírez, J. and Jiménez Osornio, J. (1988) Nescafé (*Stizolobium pruriens* [L.]) Medic. var. utilis Wall ex Wight) como un ejemplo de experimentación campesina en el trópico húmedo Mexicano. In S. del Almo ed. *Cuatro estudios sobre sistemas tradicionales*. Mexico City: Instituto Nacional Indigenista. 75–89.
Ashby, J. A. (1990) *Evaluating Technology with Farmers: a Handbook*. Cali, Colombia: Centro Internacional de Agricultural Tropical.
Bentley, J. W. (1989) What farmers don't know can't help them: The strengths and weaknesses of indigenous technical knowledge in Honduras. *Agriculture and Human Values* 6: 25–31.
Bentley, J. W. (1994) Facts, fantasies, and failures of farmer participatory research. *Agriculture and Human Values* 11: 140–150.
Buckles, D. (1995) Velvetbean: A 'new' plant with a history. *Economic Botany* 49 (1): 13–25.
Bunch, R. (1990) *Low input soil restoration in Honduras: The Cantarranas farmer-to-farmer extension programme*. Gatekeeper Series 23: 1–12.
Burkill, I. H. (1935) *A Dictionary of the Economic Products of the Malay Peninsula*. London: Crown Agents for the Colonies.

Byerlee, D. (1993) Technology adaptation and adoption: The experience of seed-fertilizer technology and beyond. *Review of Marketing and Agricultural Economics.* 6: 1–23.

Carter, E. W. (1969) *New Lands and Old Traditions: Ketchi cultivators in the Guatamelan Lowlands.* Gainsville, Florida: University of Florida Press.

Chambers, R. (1990) *Microenvironments unobserved.* International Institute for Environment and Development 22: 1–18.

Chevalier, J. M. and Buckles, D. (1995) *A Land Without Gods: Process Theory, Maldevelopment, and the Mexican Nahuas.* Halifax/London: Fernwood Books/Zed Books.

Flores, M. (1994) The use of leguminous cover crops in traditional farming systems in Central America. In D. Thurston, M. Smith, G. Abawi and S. Kearl eds. *Tapado Slash/Mulch: How Farmers Use It and What Researchers Know about it.* Ithaca, New York: Centro Agronómico Tropical de Investigación y Enseñanza and Cornell International Institute for Food, Agriculture, and Development. 149–156.

Fujisaka, S. (1989) The need to build upon farmer practice and knowledge: Reminders from selected upland conservation projects and policies. *Agroforestry Systems* 9(2): 141–153.

Graf, W., Voss, J. and Nyabenda, P. (1991) Climbing bean introduction in Southern Rwanda. In R. Tripp, ed., *Planned Change in Farming Systems: Progress in On-Farm Research.* Chichester: John Wiley. 39–62.

Granados Alvarez, N. (1989) La rotación con leguminosas como alternativa para reducir el daño causado por fitopatógenos del suelo y elevar la productividad del agrosistema maíz en el trópico humedo. MSc thesis. Montecillos, Edo. de Mexico: Colegio de Postgraduados.

Harrington, L. W., Hobbs, P. R., Tamang, D. B., Adhikari, C., Gyawali, B. K., Pradhan, G., Batsa, B., Ranjit, J. D., Ruckstuhl, M., Khadka, Y. G. and Baidya, H. L. (1992) *Wheat and Rice in the Hills: Farming Systems, Production Techniques, and Research Issues for Rice-Wheat Cropping Patterns in the Mid-Hills of Nepal.* Bangkok, Thailand: Nepal Agricultural Research Council and International Maize and Wheat Improvement Center.

Lightfoot. C. (1987) Indigenous research and on-farm trials. *Agricultural Administration and Extension* 24: 79–89.

Lightfoot, C., De Guia, O. J. and Ocado, F. (1988) A participatory method for systems-problem research: Rehabilitating marginal uplands in the Philippines. *Experimental Agriculture* 24: 301–309.

Narvaez Carvajal, G. and Paredes Hernandez, E. (1994) *El Pica-Pica (Mucuna pruriens): Mas que un abono verde para maíz, en el norte del Istmo Oaxaqueno.* Oaxaca de Juarez, Mexico: Universidad Autónoma Chapingo, Dirección de Centros Regionales.

Norman, D., Baker, D., Heinrich, G., Jonas, C., Maskiara, S. and Worman, F. (1989) Farmer groups for technology development: Experience in Botswana. In R. Chambers, A. Pacey, and L.A. Thrupp eds. *Farmer First: Farmer innovation and agricultural research.* London: Intermediate Technology Publications. 136–146.

Pare, L., Blanco R., Buckles, D., Chevalier, J. M., Gutierrez Martinez, R., Hernandez D., Perales Rivera, H. R., Ramirez R. and Velazquez H., (1993) *La Sierra de Santa Marta: Hacia un desarrollo sustentable.* Unpublished MS.

Perales Rivera, H. R. (1992) El autoconsumo en la agricultura de los Popolucas de Soteapan. Veracruz, MSc. thesis. Montecillo, Edo. de Mexico: Colegio de Postgraduados.

Perales Rivera, H. R. and Buckles, D. (1991) Experimentación campesina con el bejuco de abono (*Mucuna sp.*) en la zona indígena de la Sierra de Santa Marta. Veracruz: Una propuesta de investigación. Unpublished MS.

Scott, J. M. (1919) *Velvet Bean Varieties.* University of Florida Agricultural Experiment Station, Bulletin 152.

Stuart, J. W. (1978) Subsistence ecology of the Isthmus Nahuat Indians of Southern Veracruz, Mexico. PhD. dissertation. Riverside, California: University of California, Riverside.

Tracy, S. M. and Coe, H. S. (1918) Velvetbeans. Farmers' Bulletin 952. Washington, D.C.: United States Department of Agriculture.

Triomphe, B. (1995) Fertilidad de los suelos en la rotación maíz/mucuna, Costa Norte de Honduras: Resultados preliminares. Paper presented at the XLI PCCMCA Annual meeting, 26 March-1 April 1995, Tegucigalpa, Honduras.
Wilmot-Dear, C. M. (1984) A revision of Mucuna (Leguminosae-Phaseolease) in China and Japan. *Kew Bulletin* 39: 23–65.
Wilmot-Dear, C. M. (1987) A version of Mucuna (Leguminosae-Phaseoleae) in the Indian subcontinent and Burma. *Kew Bulletin* 42: 23–46.
Zea, J. L. (1991) Efecto residual de intercalar leguminosas sobre el rendimiento de maíz (*Zea mays* L.) en nueve localidades de Centro America. In *Sintesis de resultados experimentales*, 1991, Guatemala City, Guatemala: Programa Regional de Maíz para Centro America y el Caribe. 97–103.

Chapter 3

Chaudhary R. C. and Fujisaka S. (1992) Farmer-participatory rainfed lowland rice varietal testing in Cambodia. *Intern Rice Res Newsl 17*: 4–17.
Fujisaka S. (1990) Rainfed lowland rice: building research on farmer practice and technical knowledge. *Agric Ecosyst Environ 33*: 57–74.
Fujisaka S. (1991) A diagnostic survey of shifting cultivation in northern Laos: targeting research to improve sustainability and productivity. *Agrofor Syst 13*: 95–109.
Fujisaka S. (1993) A case of farmer adaptation and adoption of contour hedgerows for soil conservation. *Experiment Agric 29*: 97–105.
Fujisaka S., Ingram, K. T. and Moody, K. (1991a) Crop establishment (*beusani*) in Cuttack District, India. *International Rice Research Institute (IRRI) Res Paper Series 148*. IRRI: Los Baños Philippines.
Fujisaka, S., Kirk, G., Litsinger, J. A., Moody, K., Hosen, N., Yusef, A., Nurdin, F., Naim, T., Artati, F., Aziz, A., Khatib, W., and Yustisia (1991b) Wild pigs, poor soils, and upland rice; a diagnostic survey of Sitiung, Sumatra, Indonesia. *International Rice Research Institute (IRRI) Res Paper Series 155*. IRRI: Los Baños Philippines.
Fujisaka, S., Moody, K. and survey team. (1992) Rainfed lowland, deepwater, and upland rice in Myanmar: a diagnostic survey. *Myanmar J Agric Sci 4*: 1–13.
Haugerud, A. and Collinson, M. P. (1990) Plants, genes and people: improving the relevance of plant breeding in Africa. *Experiment Agric 26*: 341–62.
IRRI (International Rice Research Institute). (1988) *Agricultural economics database*. IRRI Los Baños Philippines.
IRRI (International Rice Research Institute). (1989) *IRRI toward 2000 and beyond*. IRRI: Los Baños Philippines.
Maurya, D. M., Bottrall, A. and Farrington, J. (1988) Improved livelihoods, genetic diversity, and farmer participation: a strategy for rice breeding in rainfed areas of India. *Experiment Agric 24*: 311–20.
Richards P. (1986) *Coping with hunger*. Allen and Unwin: London.
Simmonds N. W. (1991) Selection for local adaptation in a plant breeding programme. *Theor Appl Genet 82*: 363–67.
Simmonds, N. W. and Talbot, M. (1992) Analysis of on-farm rice yield data from India. *Experiment Agric 28*: 325–329.

Chapter 4

Ausenda, G. (1987) 'Leisurely Nomads: The Hadendowa (Beja) of the Gash Delta and Their Transition to Sedentary Village Life', Ph.D. thesis, Columbia University.
Baker, S. W. (1867) *The Nile Tributaries of Abyssinia, and the Sword Hunters of the Hamran Arabs*. London: Macmillan and Co.
Brouwers, J. (1994) 'Réaction des Paysans Adja à la Baisse de Fertilité du Sol', in P. Ton and L. De Haan (eds.) *A la Recherche de l'Agriculture Durable au Bénin*, pp. 55–60. Amsterdam: Instituut voor Sociale Geografie, Universiteit van Amsterdam.
Craig, G. M. (ed.) (1991) *The Agriculture of the Sudan*. Oxford: Oxford University Press.

Critchley, W., C. Reij and A. Seznec (1992) *Water Harvesting for Plant Production; Part II: Case Studies and Conclusions for Sub-Saharan Africa.* (World Bank Technical Paper No. 157), World Bank.

Dangbégnon, C. (1994) 'L'Importance des Connaissances Endogènes pour une Agriculture Durable au Sud-Bénin', in P. Ton and L. De Haan (eds.) *A la Recherche de l'Agriculture Durable au Bénin*, pp. 61–64. Amsterdam: Instituut voor Sociale Geografie, Universiteit van Amsterdam.

Hassan Mohammed Salih (1980) 'Hadanduwa Traditional Territorial Rights and Inter-population Relations Within the Context of the Native Administration System (1927–1970)'. *Sudan Notes and Records* 61: 118–33.

Hillel, D. (1992) *Out of the Earth: Civilization and the Life of the Soil.* Berkeley: University of California Press.

Hjort af Ornäs, A. and G. Dahl (1991) *Responsible Man; the Atmaan Beja of North-eastern Sudan.* Uppsala: Stockholm Studies in Social Anthropology and Nordiska Afrikainstitutet.

Jefferson, J. H. K. (1949) 'The Sudan's Grain Supply'. *Sudan Notes and Records* 30(1): 77–100.

Johnson, A. W. (1972) 'Individuality and Experimentation in Traditional Agriculture'. *Human Ecology* 1(2): 149–59.

Leakey, L. S. B. (1936) *Kenya: Contrasts and Problems.* London: Methuen.

Lebon, J. H. G. (1965) *Land Use in Sudan.* London: Geographical Publications Limited.

McLoughlin, P. F. M. (1966) 'Labour Market Conditions and Wages in the Gash and Tokar Deltas, 1900–1955'. *Sudan Notes and Records* 47: 111–126.

Newbold, D. (1935) 'The Beja Tribes of the Red Sea Hinterland', in J. A. d. C. Hamilton (ed.) *The Anglo-Egyptian Sudan from within*, pp. 140–64. London: Faber & Faber Limited.

Niemeijer, D. (1993) *Indigenous Runoff Farming in a Changing Environment: The Case of Kassala's Border Area, Sudan.* Landscape and Environmental Research Group, University of Amsterdam and Department of Irrigation and Soil and Water Conservation, Wageningen Agricultural University.

Niemeijer, D. (1996) 'The Dynamics of African Agricultural History: Is it Time for a New Development Paradigm?' *Development and Change* 27(1): 87–110.

Niemeijer, D. (1998) 'Soil Nutrient Harvesting in Indigenous Teras Water Harvesting in Eastern Sudan'. *Land Degradation & Development* 9 (4): 323–30.

Östberg, W. (1995) *Land is Coming Up: The Burunge of Central Tanzania and Their Environments.* Stockholm: Almqvist & Wiksell International.

Palmisano, A. L. (1991) *Ethnicity: The Beja as Representation.* Berlin: Das Arabische Buch.

Paul, A. (1954) *A History of the Beja Tribes of the Sudan.* Cambridge: Cambridge University Press.

Reij, C. (1991) *Indigenous Soil and Water Conservation in Africa.* London: International Institute for Environment and Development.

Reij, C., P. Mulder and L. Begemann (1988) *Water Harvesting for Plant Production.* (World Bank Technical Paper No. 91), World Bank.

Rhoades, R. and A. Bebbington (1988) 'Farmers Who Experiment: An Untapped Resource for Agricultural Research and Development', Paper presented at the International Congress on Plant Physiology, New Delhi, India.

Richards, P. (1985) *Indigenous Agricultural Revolution: Ecology and Food Production in West Africa.* London: Hutchinson.

Sørbø, G. M. (1991) 'Systems of Pastoral and Agricultural Production in Eastern Sudan', in G. M. Craig (ed.) *The agriculture of the Sudan*, pp. 214–29. Oxford: Oxford University Press.

Tothill, J. D. (ed.) (1948) *Agriculture in the Sudan.* Oxford: Oxford University Press.

Van Dam, A. J. and J. A. Houtkamp (1992) 'Livelihood and Runoff Farming: the Case of the Border Area Village of Hafarat, Eastern Region, Sudan', Master's thesis, University of Amsterdam.

Van Dijk, J. A. (1991) *Water Spreading by Broad-based Earth Embankments. WARK Final Report: Technical Performances and Socio-Economic Aspects in the Border Area, Eastern Region, Sudan*. National Council for Research & The Ford Foundation.
Van Dijk, J. A. (1995) 'Taking the Waters: Soil and Water Conservation among Settling Beja Nomads in Eastern Sudan', Ph.D. thesis, University of Amsterdam.
Van Dijk, J. A. and Mohamed Hassan Ahmed (1993) *Opportunities for Expanding Water Harvesting in Sub-Saharan Africa: The Case of the Teras of Kassala*. London: International Institute for Environment and Development.
Westoff, L. (1985) 'Socio-economic Development in Eastern Sudan: a Case Study of the Kassala Rural District', Master's thesis, University of Utrecht.

Chapter 5

Adepetu, A. A., (1985) Farmers and their farms on four fadamas on the Jos Plateau. *Jos Plateau Environmental Resources Development Programme Interim Report, No. 2*, Durham, University of Durham.
Ake, C., (1987) *Sustaining Development on the Indigenous. Long-Term Perspectives Study.* Special Economics Office. Washington, D.C. The World Bank.
Atteh, O. D., (1992) Indigenous local knowledge as key to local level development: possibilities, constraints and planning issues in the context of Africa. *Studies in Technology and Social Change*, No. 20, Technology and Social Change Program, Iowa State University, Ames.
Chambers, R., (1983) *Rural Development: Putting the Last First*. Harlow: Longman.
Chambers, R., (1989) Reversals, institutions and change. In R. Chambers, A. Pacey and L. A. Thrupp, (Eds.), *Farmer First: Farmer Innovation and Agricultural Research*. London: Intermediate Technology Publications.
Cottingham, R., (1988) Dry-season gardening projects, Niger. In C. Conroy and M. Litvinoff (Eds.), *The Greening of Aid*. London: Earthscan, pp. 69–73.
IDRC, (1993) Background to the international symposium on indigenous knowledge and sustainable development. *Indigenous Knowledge and Development Monitor*. 1, (2), 2–5.
Morgan, W. T. W., (Ed.) (1979) The Jos Plateau: a survey of environment and land use. *Occasional Publications (New Series), No. 14*, Durham, Department of Geography, University of Durham.
Morgan, W. T. W., (1985) Forward. In farmers and their farms on four fadamas on the Jos Plateau, by A. A. Adepetu. *Jos Plateau Environmental Resources Development Programme Interim Report, No. 2*, Durham, University of Durham.
Phillips, A. O., (1990) *Economic Impact of Nigeria's Structural Adjustment Programme*. NISER Monograph Series No. 1. Ibadan, NISER.
Phillips-Howard, K. D., A. A. Adepetu and A. D. Kidd, (1990). Aspects of change in fadama farming along the Delimi River, Jos L. G. C. (1982–1990). *Jos Plateau Environmental Resources Development Programme Interim Report, No. 18*, Durham, University of Durham.
Phillips-Howard, K. D. and A. D. Kidd, (1990) Significance of indigenous knowledge systems to the improvement of dry-season farming on the Jos Plateau. *Jos Plateau Environmental Resources Development Programme Interim Report, No. 17*, Durham, University of Durham.
Phillips-Howard, K. D. and A. D. Kidd, (1991) Knowledge and management of soil fertility among dry-season farmers on the Jos Plateau, Nigeria. *Jos Plateau Environmental Resources Development Programme Interim Report, No. 25*, Durham, University of Durham.
Porter, G. and K. D. Phillips-Howard, (1993) Bitter brew? Barley growing in Nigeria. *Geography Review, 6*, (5), 34–37.
Slikkerveer, L. J., (1993) 'Transformation-of-technology and indigenous approaches to innovation: towards a new paradigm of sustainable development.' Paper presented at the Pithecanthropus Centennial 1893–1993, 'Human Evolution in its Ecological Context', held in Leiden, The Netherlands, 26 June-1 July 1993.

Titilola, S. O., (1992) 'The role of indigenous knowledge in rural development activities for sustainable development in Nigeria.' Paper presented at the International Symposium on Indigenous Knowledge and Sustainable Development, held at the International Institute for Rural Reconstruction, Silang, Cavite, Philippines, 21–25 September, 1992.

Chapter 6

Altieri, M. A. and Merrick, L. C. (1988) Agroecology and *In Situ* Conservation of Native Crop Diversity in the Third World. In Wilson, E. O. (ed) *Biodiversity*. Washington: National Academy Press.

Boncodin, R. and G. Prain (1997) The Dynamics of Biodiversity Conservation by Homegardeners in Bukidnon, Southern Philippines. In *Local R&D: Institutionalizing and sustaining agricultural innovation in Asia*. Los Baños: UPWARD.

Brush, S. (1991) A farmer-based approach to conserving crop germplasm. *Economic Botany* 45(2) 153–65.

Brush, S. (1992) Ethnoecology, biodiversity, and modernization in Andean Potato Agriculture. *J. Ethnobiol.* 12(2): 161–85.

Brush, S. (1993) 'In situ conservation of landraces in Centers of Crop Diversity.' Paper presented for the Symposium on Global Implications of Germplasm conservation and utilization at the 85th Annual Meetings of the American Society of Agronomy, Cincinnati, OH.

Brush, S. B., H. J. Camey and Z. Huaman (1981) 'Dynamics of Andean potato agriculture'. *Economic Botany* 35(1): 70–88.

Chambers, R. (1992) *Rural Appraisal. Relaxed and Participatory*. IDS Discussion paper No. 331 October, Brighton, Sussex Institute of Development Studies.

Castillo, G. T. (1995) Secondary crops in primary functions: The search for systems, synergy, and sustainability. UPWARD Working Paper Series #2. Los Baños, Laguna: UPWARD.

Edgerton, Ronald K. (1982) Frontier Society on the Bukidnon Plateau. In McCoy, A. W. and Ed. C. de Jesus (eds.) *Philippine Social History: Global Trade and Local Transformations*. Manila: Ateneo de Manila University Press.

Geertz, C. (1963) *Agricultural involution. The processes of ecological change in Indonesia*. London: University of California Press.

Ingold, T. (1993) Globes and spheres: the topology of environmentalism in Miltan, K. (ed). *Environmentalism: the view from anthropology*. ASA Monographs 32 London: Routledge.

IPGRI (1993) *Diversity for Development. The Strategy of the International Plant Genetic Resources Institute*. Rome: International Plant Genetic Resources Institute.

Joshi, A and J. R. Whitcombe (1996) Farmer Participatory Crop Improvement. II. Participatory Varietal Selection, a case study in India. *Experimental Agriculture* Vol. 32, pp. 461–77.

Nazarea-Sandoval, V. (1991) Memory Banking of Indigenous Technologies Associated with Traditional Crop Varieties: a focus on sweet potatoes. In *Sweetpotato Cultures of Asia and South Pacific*. Proceedings of the 2nd. Annual UPWARD International Conference. Los Baños, Philippines.

Nazarea-Sadoval, V. (1994a) Memory banking: The conservation of cultural and genetic diversity in sweet potato production. In: Prain, G. D. and C. Bagalanon (eds). *Local knowledge, global science and plant genetic resources: towards a partnership*. Proceedings of the International Workshop on Genetic Resources. Los Baños, Laguna: UPWARD.

Nazarea-Sandoval, V. (1994b) Memory banking protocol: A guide for documenting indigenous knowledge associated with traditional crop varieties. In Prain, G. D. and C. Bagalanon (eds). *Local knowledge, global science and plant genetic resources: towards a partnership*. Proceedings of the International Workshop on Genetic Resources. Los Baños, Laguna: UPWARD.

Prain, G. (1995) Sweet potato in Asian Production Systems: an overview of UPWARD's first phase research. In *Taking Root*: Proceedings of the Third UPWARD Review and Planning Workshop. Los Baños: UPWARD.

Prain, G., Il Gin Mok, T. Sawor, P. Chadikun, E. Atmodjo and E. Relwaty Sitmorang (1995) Interdisciplinary collecting of *Ipomoea batatas* germplasm and associated indigenous knowledge in Irian Jaya. In Guarino, L., V. Ramanatha Rao and R. Reid (eds), *Collecting Plant Genetic Diversity.* Oxford: CAB International.

Prain, G. and M. Piniero (1998) Communal Conservation of Rootcrop Genetic Resources in Southern Philippines. In Prain, G. and C. Bagalanon (eds) *Conservation and Change: farmer management of agro-biological diversity in the Philippines.* Los Baños: UPWARD.

Sthapit, B. R., K. D. Joshi and J. R. Whitcombe (1996) Farmer Participatory Crop Improvement. III. Participatory Plant Breeding, a case study for rice in Nepal. *Experimental Agriculture* Vol. 32, pp. 479–96.

Whitcombe, J. R., A. Joshi, K. D. Joshi and B. R. Sthapit (1996) Farmer Participatory Crop Improvement. I. Varietal Selection and Breeding Methods and their Impact on Biodiversity. *Experimental Agriculture* Vol. 32, pp. 445–60.

Yen, D. (1974) *The Sweet potato and Oceania. An essay in Ethnobotany.* Honolulu, Hawaii: Bishop Museum Press.

Vega, Amihan Belita and José L. Bawsmo (1998) Community-based Sweet Potato Genebanking and Distribution System. In Prain, G. and C. Bagalanan (eds) *Conservation and Change: farmer management of agro-biological diversity in the Philipplies.* Los Baños: UPWARD.

Chapter 7

Altieri, M. (1987) 'The significance of diversity in the maintenance of the sustainability of traditional agroecosystem.' *ILEIA Newsletter* (2): 3–7.

Badri, B. and A. Badri. (1994) 'Women and Biodiversity.' *Development* 1: 67–71.

Bentley, J. W. (1990) 'Conocimiento y experimentos espontáneos de campesinos hondureños sobre el maíz muerto' [Honduran peasants' knowledge and spontaneous experiments about dead-maize]. *Manejo Integrado de Plagas (Costa Rica)* 17: 16–26.

Berkes, F. and C. Folke. (1994) 'Linking social and ecological systems for resilience and sustainability.' Paper presented at the Workshop on 'Property Rights and the Performance of Natural Resource Systems', Stockholm (Sweden): The Beijer International Institute of Ecological Economics, The Royal Swedish Academy of Sciences.

Chambers, R. (1994) 'Afterword', pp. 264–65 in I. Scoones and J. Thompson (eds). *Beyond Farmer First. Rural people's knowledge, agricultural research and extension practice.* London: Intermediate Technology Publications.

Cunningham, A. B. (1991) 'Indigenous Knowledge and Biodiversity. "Global commons or regional heritage?"' *Cultural Survival Quarterly* 15(2): 4–8.

Gliessman, S. R., R. Garcia and M. Amador (1981). 'The ecological basis for the application of traditional agricultural technology in the management of tropical agro-ecosystems.' *Agro-ecosystems* 7(3): 173–175.

Haverkort, B. W. Himestra, C. Rejintjes and S. Essers. (1988) 'Strengthening Farmers' Capacity for Technology Development.' *ILEIA Newsletter* 4(3): 3–7.

Haverkort, B. and D. Millar. (1992) 'Farmers' experiments and cosmovision.' *ILEIA Newsletter* 8(1): 26–27.

Haverkort, B. and D. Millar. (1994) 'Constructing Diversity: The Active Role of Rural People in Maintaining and Enhancing Biodiversity.' *Etnoecologica* 2(3): 51–63.

Hyndman, D., (1992) 'Ancient Futures for Indigenous People: Cultural and Biological Diversity through Self-Determination.' Paper presented at the International Symposium on 'Indigenous Knowledge and Sustainable Development'. Silang, Cavite (The Philippines): IIRR.

Jiggins, J. (1994) *Changing the Boundaries. Women-centered perspectives on Population and the Environment.* Washington D.C.: Island Press.

Johnson, A. W. (1972) 'Individuality and Experimentation in Traditional Agriculture.' *Human Ecology* 1(2): 149–59.
Mathias, E. (1994) *Indigenous Knowledge and Sustainable Development*. International Institute of Rural Reconstruction Working paper No. 53. Silang, Cavite (The Philippines): IIRR.
McCorkle, C. (1994) *Farmer Innovations in Niger*. Studies in Technology and Social Change Series, No. 21. Technology and Social Change Program in Collaboration with Center for Indigenous Knowledge for Agriculture and Rural Development (CIKARD), Ames Iowa: Iowa State University.
Norem, R. H., R. Yoder and Y. Martin. (1989) 'Indigenous Agricultural Knowledge and Gender Issues in Third World Agricultural Development', pp. 91–100 in D. M. Warren, L. J. Slikkerver and S. O. Tilola (eds). *Indigenous Knowledge Systems: Implications for Agriculture and International Development*. Studies in Technology and Social Change program No. 11. Ames, Iowa: Iowa State University.
Quiroz, C. (1992) 'The use of the Interpretive Research Approach as a tool for understanding Farmers' Agricultural Indigenous Knowledge Systems: An example from Venezuela.' Paper presented at the International Symposium on 'Indigenous Knowledge and Sustainable Development'. Silang, Cavite (The Philippines): IIRR.
Quiroz, C. (1994) 'Biodiversity, Indigenous Knowledge, Gender and Intellectual Property Rights.' *Indigenous Knowledge & Development Monitor* 2(3): 12–15.
Reijntjes, C. and W. Hiemstra. (1989) 'Farmer Experimentation and Communication? *ILEIA Newsletter* 5(1) 3–6.
Rengifo-Vasquez, R. (1989) 'Experimentación campesina' [Peasants' Experimentation]. *Documento de Estudio* No. 16, Lima, Peru: PRATEC.
Rhoades, R. and A. Bebbington. (1988) 'Farmers who experiment: an untapped resource for agricultural resarch and development.' Paper presented at the International Congress of Plant Physiology in New Delhi, India.
Rocheleau, D. E. (1991) 'Gender, Ecology, and the Science of Survival: Stories and Lessons from Kenya? *Agriculture and Human Values* 8(1/2): 156–65.
Scoones, I and J. Thompson. (1994) 'Knowledge, power and agriculture – towards theoretical understanding', in I. Scoones and J. Thompson (eds). *Beyond Farmer First: Rural People's Knowledge, Agricultural Research and Extension Practice*. London: Intermediate Technology Publications. 16–32.
Shiva, V. (1993) *Monocultures of the Mind: Perspectives on Biodiversity and Biotechnology*. London and New Jersey: Zed Books/Third World Network.
Shiva, V. and I. Dankelman. (1992) 'Women and Biological Diversity: Lessons from the Indian Himalaya: in D. Cooper, R. Vellue and H. Hobbelink (eds). *Growing Diversity: Genetic Resources and Local Food Security*. London: Intermediate Technology Publications. 44–50.
Stolzenback, A. (1993) 'Farmers' experimentation: what are we talking about?' *ILEIA Newsletter* 9(1): 28–29.
UNEP. (1994) 'Intergovernment Committee on the Convention of Biological Diversity', Second Session, UNDEP/CBD/IC/2/14, Nairobi: UNEP.
Warren, D. M. (1991) *Using Indigenous Knowledge in Agricultural Development*. World Bank Discussion Papers No. 127, Washington, D.C.: The World Bank.
Warren, D. M. (1992) 'Indigenous Knowledge, Biodiversity Conservation and Development'. Keynote Address presented at the International Conference on 'Conservation of Biodiversity in Africa: Local Initiatives and Institutional Roles'. Nairobi, Kenya: National Museums of Kenya.

Chapter 9

Altieri, M. A. and L. C. Merrick. (1987) *In situ* conservation of crop genetic resources through maintenance of traditional farming systems. *Economic Botany* 41 (1): 86–96.
Cleveland, D. and D. Soleri. (1987) Household gardens as a development strategy. *Human Organization* 46 (3): 259–70.

Dei, G. J. S. (1988) Crisis and adaptation in a Ghanaian forest community. *Anthropology Quarterly* 62 (1): 63–72.

Dei, G. J. S. (1989) Hunting and gathering in a Ghanaian rain forest community. *Ecology of Food and Nutrition* 22 (3): 225–44.

Gomez-Pompa, A. (1987) On Maya silviculture. *Mexican Studies* 3 (1): 1–17.

Grubben, G. H. J. (1977) *Tropical vegetables and their genetic resources.* Food and Agriculture Organization, Rome: International Board for Plant Genetic Resources.

Harlan, J. (1975) *Crops and Man.* Madison, Wisconsin: American Society of Agronomy and Crop Science Society of America.

Hynes, R. A. and A. K. Chase. (1982) Plant sites and domiculture: Aboriginal influence upon plant communities in Cape York Peninsula. *Archaeological Oceania* 17: 38–50.

Kumar, S. K. (1978) *Role of the household economy in child nutrition at low-incomes: A case study in Kerala.* Occasional Paper No. 95. Ithaca, New York: Department of Agricultural Economics, Cornell University.

Kundstadter, P. (1978) Ecological modification and adaptation: An ethnobotanical view of Lua Swiddeners in northwestern Thailand. *Anthropological Papers.* 67: 169–200.

Mwajumwa, L. B. S., E. M. Kahangi, and J. K. Imung. (1991) The prevalence and nutritional values of some Kenyan indigenous leafy vegetables from three locations of Machakos district. *Ecology of Food and Nutrition* 26: 275–80.

Ninez, V. (1987) Household gardens: Theoretical and policy considerations. *Agricultural Systems* 23: 167–86.

Posey, D. A. (1990) The science of the Mebengokre. *Orion Summer* 9 (3): 16–23.

Rajasekaran, B. (1994) *A framework for incorporating indigenous knowledge systems into agricultural research and extension organizations for sustainable agricultural development.* Technology and Social Change Series, No. 22. Ames, Iowa: Technology and Social Change Program, Iowa State University.

Rajasekaran, B. and M. B. Whiteford. (1993) Rice-crab production: The role of indigenous knowledge in designing food security policies. *Food Policy* 18 (3): 237–47.

Rajasekaran, B. and D. M. Warren. (1994) Using indigenous knowledge as a tool for socioeconomic development and biodiversity conservation in the hilly regions of India: The case of Kolli hills. *Indigenous Knowledge and Development Monitor* 2 (4): 13–17.

Richards, P. (1985) *Indigenous Agricultural Revolution: Ecology and Food Production in West Africa.* London: Hutchinson.

Roling, N. and P. Engel. (1992) The development and concept of agricultural knowledge information systems (AKIS): Implications for Extension. In W. M. Rivera and D. J. Gustafson (eds.), *Agricultural Extension: Worldwide Institutional Evolution and Forces for Change.* New York: Elsevier Science Publishing Company. 125–38.

Roosevelt, A. (1990) The historical perspective of resource use in tropical Latin America. *Economic Catalysts to Ecological Change.* Gainesville, Florida: Center for Latin American Studies, University of Florida. 30–64.

Soemarwoto, O. and G. R. Conway. (1991) Javanese home gardens. *Journal for Farming Systems Research-Extension* 2 (3): 95–118.

Soemarwoto, O. and I. Soemarwoto. (1984) The Javanese rural ecosystem. In A. T. Rambo and P. E. Sajise (eds.), *An Introduction to Human Ecology Research on Agricultural Systems in South East Asia,* Los Baños: University of the Philippines. 261–270.

Stoler, A. (1979) Garden Use and Household Economy in Java. In G. E. Hansen, (ed.), *Agriculture and Rural Development in Indonesia.* Boulder, Colorado: Westview Press. pp 242–54.

Turner II, B. L. and C. H. Miksicek. (1984) Economic plant species associated with prehistoric agriculture in the Maya lowlands. *Economic Botany* 38 (2): 179–93.

Thurston, H. D. (1991) *Sustainable Practices for Plant Disease Management in Traditional Farming Systems.* Boulder, Colorado: Westview.

Warren, D. M. (1991) *Using Indigenous Knowledge in Agricultural Development.* World Bank Discussion Paper No. 127. Washington, D.C.: The World Bank.

Warren, D. M. (1992) 'Indigenous knowledge, biodiversity conservation and development.' Keynote Address at International Conference on Conservation of Biodiversity in Africa: Local Initiatives and Institutional Roles. Nairobi, Kenya, August 30–September 3, 1992.

Whiteford, M. B. (1995) *Como Se Cura*: Patterns of medical choice among working class families in the city of Oaxaca, Mexico. In D. M. Warren, L. J. Slikkerveer and D. Brokensha (eds.), *Indigenous Knowledge Systems: The Cultural Dimensions of Development*. London: Intermediate Technology Publications. 218–230.

Chapter 10

Berger, P. L., and T. Luckmann (1967) *The social construction of reality. A treatise in the Sociology of Knowledge*. New York: Anchor Books.

Brouwers, J. H. A. M. (1993) *Rural people's response to soil fertility decline, The Adja case (Benin)*. Wageningen Agricultural University Paper 93–4.

Campbell, A. (1994) *Landcare. Communities shaping the land and the future.* St Leonards (Australia): Allen and Unwin.

Checkland, P. (1981) *Systems Thinking, Systems Practice.* Chichester: John Wiley.

Checkland, P. and J. Scholes. (1990) *Soft Systems Methodology in Action.* Chichester: John Wiley.

Dangbégnon, C., and J. H. A. M.Brouwers. (1990) Maize farmers' informal Research and Development. *ILEIA newsletter* 90(3): 24–25.

Dangbégnon, C., and J. H. A. M.Brouwers. (1991) *Réseau d'informateurs: une expérience pour l'identification des connaissances locales sur le plateau Adja.* Report Agricultural Faculty, Beninese National University, Cotonou.

Darré, J. P. (1985) *La Parole et la Technique. L'univers de pensée des éleveurs du Ternois.* Paris: l'Harmattan.

Giddens, A. (1987) *Social Theory and Modern Sociology.* Cambridge: Polity Press.

Giddens, A. (1990) 'Jurgen Habermas'. Ch. 7 in: Skinner I. Q. (Ed.) *The return of grand theory in the human sciences.* 6th ed., Cambridge: Cambridge University Press.

Haverkort, A., J. Van der Kamp, and A. Waters-Bayer (Eds.). (1991) *Joining farmers' experiments. Experiences in Participatory Technology Development.* London: Intermediate Technology Publications.

Jiggins, J. L. S. and H. De Zeeuw. (1992) Participatory technology development in practice: process and methods. In: C. Reijntjes, B. Haverkort and A. Waters-Bayer, *Farming for the Future: an introduction to low external input agriculture.* London: MacMillan and Leusden: ILEIA. 135–62.

Kline, S., and N. Rosenberg. (1986) An Overview of Innovation. In: Landau, R. and N. Rosenberg (Eds.). *The Positive Sum Strategy. Harnessing Technology for Economic Growth.* Washington (DC): National Academic Press, pp 275–306.

Koudokpon, V., J. H. A. M. Brouwers, M. N. Versteeg, A. Budelman. (in press) 'Priority setting in research for sustainable land use: The case of the Adja Plateau, Benin.' accepted by *Agroforestry Systems.*

Lionberger, H. and C. Chang. (1970) *Farm Information for Modernising Agriculture: The Taiwan System.* New York: Praeger.

Matteson, P., K. D. Gallagher, and P. E. Kenmore. (1992) Extension and integrated pest management for planthoppers in Asian irrigated rice. In Denno, R. F. and T. J. Perfect (Eds.). *Ecology and Management of Plant Hoppers.* London: Chapman and Hall.

Rogers, E. M. (1995) *Diffusion of Innovations.* (4th edition) New York: Free Press.

Röling, N. G. (1988) *Extension Science: Information systems in agricultural development.* Cambridge: Cambridge University Press.

Röling, N. G. (1993) Agricultural knowledge and environmental regulation: the Crop Protection Plan and the Koekoekspolder, *Sociologia Ruralis,* 33(2), June: 212–31.

Röling, N. G. (1994) Platforms for decision making about eco-systems. Chapter 31 of L. O. Fresco *et al.* (Eds), *Future of the Land: Mobilising and Integrating Knowledge for Land Use Options.* Chichester: John Wiley and Sons, Ltd, pp 386–93.

Röling, N. G. and P. Engel. (1989) IKS and knowledge management: utilising knowledge in institutional knowledge systems. In: Warren, D. M., L. J. Slikkeveer, and S. Oguntunji Tititola (Eds.). *Indigenous Knowledge Systems: Implications for Agriculture and International Development.* Ames (Iowa): Iowa State University: Technology and Social Change Programma, Studies in Technology and Social Change, No 11, pp 101–116.

Röling, N. G. and P. Engel. (1991) The development of the concept of Agricultural Knowledge and Information Systems (AKIS): Implications for Extension. In: Rivera, W. and D. Gustafson (Eds.). *Agricultural Extension: Worldwide Institutional Evolution and Forces for Change.* Amsterdam: Elsevier Science Publishers, pp 125–139.

Röling, N. G. and E. Van de Fliert. (1994). Transforming extension for sustainable agriculture: the case of Integrated Pest Management in rice in Indonesia. *Agriculture and Human Values* Vol 11 (2+3), Spring and Summer, pp 96–108. An adapted version of this article appeared in: Röling, N. G. and E. Van de Fliert (1998). Transforming Extension For Sustainable Agriculture: The Case of Integrated Pest Management in Rice in Indonesia. In: N. Röling and A. Wagemakers (Eds.). *Facilitating Sustainable Agriculture. Participatory Learning and Adaptive Management in Times of Environmental Uncertainty.* Cambridge: Cambridge University Press.

Somers, B. M., and N. Roling. (1993) *Ontwikkeling van kennis voor duurzame landbouw: een verkennende studie aan de hand van enkele experimentele projekten.* Den Haag: NRLO.

Van de Fliert, E. (1993) *Integrated pest management: farmer field schools generate sustainable practices. A case study in Central Java evaluating IPM training.* Wageningen: Wageningen Agricultural University Papers 93-3.

Van de Fliert, E., K. van Elsen and F. Nangsir Soenanto. (1993) Integrated Rat Management: A Community Activity. Results of a Pilot Programme in Indonesia. *FAO Plant Prot. Bull.*, 41(3).

Van der Ley, H. A. and M. D. C. Proost. (1992) *Gewasbescherming met een toekomst: de visie van agrarische ondernemers: een doelgroepverkennend onderzoek ten behoeve van voorlichting.* Wageningen: Agricultural University, Department of Communication and Innovation Studies.

Van der Ploeg, J. D. (1991) *Landbouw als mensenwerk.* Muiderberg: Countinho.

Van Kessel, J. (1988) *Aymara technologie: een culturele benadering.* Workbook for the course Sociology/Anthropology of Development Policy, Amsterdam: Free University, Dept. CASNW.

Van Weeperen, W., J. Proost and N. Röling (1998). Introducing integrated arable farming in the Netherlands. Chapter 6 in: N. Röling and A. Wagemakers (Eds.). *Facilitating Sustainable Agriculture. Participatory Learning and Adaptive Management in Times of Environmental Uncertainty.* Cambridge: Cambridge University Press.

Chapter 11

Boster, J. (1985). Selection for perceptual distinctiveness: evidence from Aguaruna cultivars of *Manihot esculenta. Economic Botany* 39: 310–25.

Bulmer, R. (1965). Beliefs concerning the propagation of new varieties of sweet potato in two New Guinea Highland societies. *Journal of the Polynesian Society* 74: 237–39.

Golson, J. and D. Gardner (1990). Agriculture and sociopolitical organization in New Guinea Highlands prehistory. *Annual Review of Anthropology* 19: 395–417.

Heider, K. (1969). Sweet potato notes and lexical queries, or, the problem of all those names for sweet potato in the New Guinea Highlands. *Kroeber Anthropological Society Papers* 41: 78–86.

Matanubun, H. *et al.* (1991). *Eksplorasi I ubi-ubian di Dabaupaten Paniai dan Kabupaten Biak-Numfor, Iran Jaya.* RTCRC Research Publications, No. 3. Manokwari: PSU UNCEN.

Schneider, J. *et al.* (1993). *Sweet potato in the Baliem Valley area, Irian Jaya: A report on collection and study of sweet potato germplasm, April-May 1993.* Bogor: CIP/CRIFC/RTCRC.

Yen, D. (1974). *The sweet potato and Oceania: An essay in ethnobotany.* Honolulu, Hawaii: Bishop Museum.

Chapter 12

Gubbels, Peter. (1988) Peasant farmer agricultural self-development. *ILEIA Newsletter* 4 (3): 11–14.

Gubbels, Peter. (1992) *Farmer-first research: Populist pipedream or practical paradigm? Prospects for indigenous agricultural development in West Africa.* MSc Dissertation, School of Development Studies, University of East Anglia.

ETC. (1991) *Learning for people-centered technology development. A training guide.* Draft version 1.1 Part 4: Farmers' experimentation. Leusden, The Netherlands.

Hoveyn, C. A. (1991) *Proeven met planten. Van opzet tot analyse.* Wageningen, CABO-DLO.

Richards, Paul. (1987) Agriculture as a performance. In: Chambers, R., A. Pacey, and L. A Thrupp (eds.), 1989: *Farmer first. Farmer innovation and agricultural research.* London, Intermediate Technology Publications.

Schön, Donald A. (1983) *The reflective practitioner. How professionals think in action.* New York, Basic Books.

Stolzenbach, Arthur F. V. (1992) *Learning by improvisation: The logic of farmers' experimentation in Mali.* Paper prepared for the IIED/IDS Beyond Farmer First: Rural People's Knowledge Agricultural Research and Extension Practice Workshop. Institute of Development Studies, University of Sussex, 27–29 October 1992.

Stolzenbach, Arthur F. V. (1993) Farmers' experimentation: what are we talking about? *ILEIA Newsletter* 9 (1): 28–29.

Van der Ploeg, Jan Douwe. (1987) *De verwetenschappelijking van de landbouwbeoefening.* Wageningen, Medelingen van de Vakgroep Sociologie, Landbouwuniversiteit, 21.

Chapter 13

Association for Rural Advancement. (1990) Maputaland: Conservation and Removals. Special Report No. 6. AFRA, Pietermaritzburg.

Begg, G. W. (1980) 'The kosi system: aspects of its biology, management and research.' In: Bruton, M. N. and Cooper, K. H. (Ed.) *Studies on the ecology of Maputaland.* Rhodes University, Grahamstown and Natal Branch of the Wildlife Society of Southern Africa, Durban.

Bryant, A. T. (1949) *The Zulu people: As they were before the white man came.* Shuter and Shooter, Pietermaritzburg.

Centre for Community Organisation Research and Development (CORD) (1991a) 'Regaining Control'. In: Ramphele, M. *Restoring the land, environment and change in post-apartheid South Africa.* Panos Publications, London.

CORD. (1991b) Overcoming apartheid's land legacy in Maputaland (Northern Natal). Working Paper 3. University of Natal, Durban.

Cleveland, D. A. and Soleri, D. (1991) *Food from dryland gardens. An ecological, nutritional and social approach to small-scale household food production.* Centre for People, Food and Environment, Tucson.

Felgate, W. D. (1982) *The Tembe Thonga of Natal and Mozambique: An ecological approach.* (Edited and arranged by E. J. Krige) Department of African Studies. University of Natal, Durban.

Granger, J. E., Hall, M., Mckenzie, B. and Feely, J. M. (1985) Archaeological research on plant and animal husbandry in Transkei. *South African Journal of Science.* Vol. 81. 12–15.

Junod, H. A. (1927) *The life of a South African Tribe.* Vols, 1 & 2. MacMillan, London.

Kwazulu Government (1992) Draft white paper on development policy in the Ubombo/ Ingwavuma Region.

Macvicar, C. N., De Villers, J. M., Loxton, R. F., Verster, E. *et al.* (1977) *Soil classification.*

www.ingramcontent.com/pod-product-compliance
Lightning Source LLC
Jackson TN
JSHW011351130125
77033JS00016B/566
9781853394430